Really ?

Images courtesy Pixabay

Images by Kira Borkia   and   Jenskuvain from Pixabay

The Extraterrestrials Are Here
Who knows What, from Where, or How ?

Acknowledgments

I am grateful to many for much. Here are a few:

I wish to thank NASA for data and beautiful pictures.

I wish to thank Space.com for data and wonderful pictures.

I wish to thank PIXABAY and its members for fantastic images.

I wish to thank Kanterbury Karen for proofreading and encouragement.

I wish to thank readers, and I hope that this book entertains, provokes thought, and teaches a bit of physics in a painless and fun way.

I wish to thank my wonderful wife, Muriel, for her gracious patience as I robotically ignored her for so many writing hours each day.

A robot author

at work

and 'his' wife

Image courtesy Sabrina Belle of PIXABAY

Other books by Edmund here --> www.amazon.com/author/edmundjgoodwinsbooks

The Extraterrestrials Are Here
They're not Earth People, but...

Suppose that you believed there were other intelligent life forms on other planets. Then, suppose that some of these non-Earthly people from other planets, somewhere in the universe, were capable of traveling to Earth. What questions would you ask yourself in order to solidify your belief ?

Before you begin to build a list of such questions, you might research some of the huge plethora of evidence that your belief is actually true. The 'Sherlock Holmes Method' is a very fine and logical way to reach the truth. That mythical character said that, if you eliminate **_all_** other possibilities, then one of the remaining possibilities must be true. If there is only one possibility left, then it is the true one. Note the stress on the word: **_all_**. One cannot leave any possibility off of your list ! Furthermore, **_any_** 'possibility' that breaks the laws of physics is automatically eliminated.'

Questions to be resolved are, of course, the six "Wh's" plus one: "What, Who, Where, Why, When, Which, and How". "Are they" and "Will they" are also good questions. This quest for truth will not include ridiculous ideas such as Hollow Earth civilizations or Atlantis is still alive. *kiss* Keep It Simple, Stupid ! The purpose of this book is to use laws of physics, logic and true data to look at the probability of extraterrestrials existing, then the probability of them being here on Earth already, and finally to figure out how to communicate with them, trade with them, and live with them in some manner.

I hope you enjoy the ride !

Why not ?

In all seriousness, the TV series 'Star Trek, The Next Generation' and ; 'Star Trek, Deep Space 9' are actually quite realistic in many ways, though not in all ways.
Highly recommended ! Just open your mind a bit.

Index

Chapter one – What ? *** *** *** *** **   Page 1
Chapter two – How ? *** *** *** *** ***   Page 11
Chapter three –   More Hows ? *** *** *   Page 21
Chapter four   Who ? *** *** *** ***   Page 31
Chapter five   When ? *** *** *** **   Page 38
Chapter six   Where ? *** *** ***   Page 43
Chapter seven Which ? *** *** *** *** **   Page 49
Chapter eight   Why ? *** *** *** ***   Page 55
Chapter nine   True or False ? *** ***   Page 63

"With archaic age comes decades of true history,
a large amount of knowledge, and sometimes
a small measure of wisdom-- if you're lucky !"
Uncertain author 2024

Chapter one –   What ?

What indeed ! There are many 'whats'. Here is a short, partial list:

What evidence exists that they exist ?	Page 1
What do they look like ?	Page 5
What do they eat ?	Page 8
What do they smell like ?	Page 8
What senses do they have ?	Page 8
What might they like ?	Page 8
What diseases might plague them ?	Page 9
What do they want from us, if anything ?	Page 9
What help may they offer to us Earthlings ?	Page 9
What should we worry about, if anything ?	Page 9

We will peek at each of these. Lets begin with the first question:

What evidence exists that they exist ?

Let us start with an appalling subject because it suggest that, almost certainly, intelligent, non-Earthly beings are responsible, and possibly why. A rather gruesome set of evidences is animal mutilation. There are hundreds of cases throughout the land and around the globe. Mostly, they are all very similar.

Often the animal is a cow. Most often the mutilations are very surgical in terms of the cuts. Organs are removed rather precisely. Blood has been drained completely, yet the surroundings have no blood spill. No tracks exist in the nearby dirt. No predator marks are on the corpse unless after death. In some of the cases, there are marks in the surrounding dirt of a huge circle made by something very heavy.

An important bit of information is that research shows these seemingly horrible acts have very rarely (if ever, at all) been done to humans. But occasionally... an example story below:

https://www.history.com/news/ufos-aliens-animal-human-mutilation-lovette-cunningham

The expert investigation of 'Skinwalker Ranch' by respected scientists and researchers on the History TV Channel has proof of animal mutilations. They have also discovered a huge amount of evidence, over many years, that points to extraterrestrial activity on that ranch in Northwest Utah ! They also have scientific evidence of a portal to somewhere. It is well worth watching, and the series is **not** like so many others that make unfounded, unbelievable, fantastic claims.

So, why would the gruesome subject of animal mutilation be good and reasonable proof of extraterrestrials ? If off-world civilizations sent teams (or robotic devices) to examine creatures on our planet, would they almost always avoid killing the (seemingly) intelligent beings of this world ? Yes !

Would they most often kill and examine other animals more thoroughly in order to determine what makes Earth-life tick ? Yes, of course ! They have to be very smart beings to travel such vast distances to get here. Learning the biology of Earth life is necessary for further contact in the future. We will see why later...

Whew ! Enough gore ! Now, lets consider subjects that aren't so gory. There are a very large number of photos of real UFOs (now called UAPs-- Unidentified Aerial Phenomena). Sadly, many of them are fake, making it much more difficult to generate real interest in real investigations by real scientific researchers.

This next image is a much too perfect fake:

Image courtesy PIXABAY

The following is a genuine photo that is not fake. On August 4, 1990, two hikers near Calvine, Scotland took a photograph of a mysterious, diamond-shaped flying object hovering in the middle of the sky. In the photo-- one of a series of six the hikers reportedly took-- a diamond-shaped object can be seen flying in the sky, while a fighter jet can be spotted in the background not too far from it.The six photos have been examined by many governments and many more experts for a bunch of years. Here below is a blow-up view:

The Calvine Scotland photo cropped and magnified

The whole story can be found here:

https://www.newsweek.com/best-ufo-picture-calvine-photo-found-30-years-missing-1733673

As of this writing, Congress is holding hearings on UFOs/UAPs. There are some absolutely shocking sightings by our military of cloaked vehicles diving into the ocean from 60,000 feet at speeds that are many times the speed of sound (which is around 800 mph). The object seemed to defy gravity. It would be nice if Congress also considered this bit of 1990 news as well as the newest sightings, without government censorship. We shall see...

The full image from the 1990 sighting is next:

One of the six Calvine Scotland photos

The latest examples on TV news at this time are newly released videos taken by military jets of UAPs. The new Congressional Hearing on UFOs and UAPs have shaken loose a huge number of videos and pictures from secretly hidden sources. Finally !

The new Congressional Hearing has also disclosed that the U. S. government has a team to reverse engineer alien craft technology. Apparently, just as suspected, some UFOs have crashed and are no longer Unidentified Flying Objects. We should call them by their proper names: Extraterrestrial Spacecrafts !

Some images and information are easily available here:

https://www.archives.gov/research/topics/uaps

This next picture is a snap-shot-part of a quite long targeting video of a typical one of them and it is not a fake. The video it came from was taken twenty years ago and hidden in secret until recently. The vehicle is not ours and cannot be from another country either. It was traveling at a very high speed:

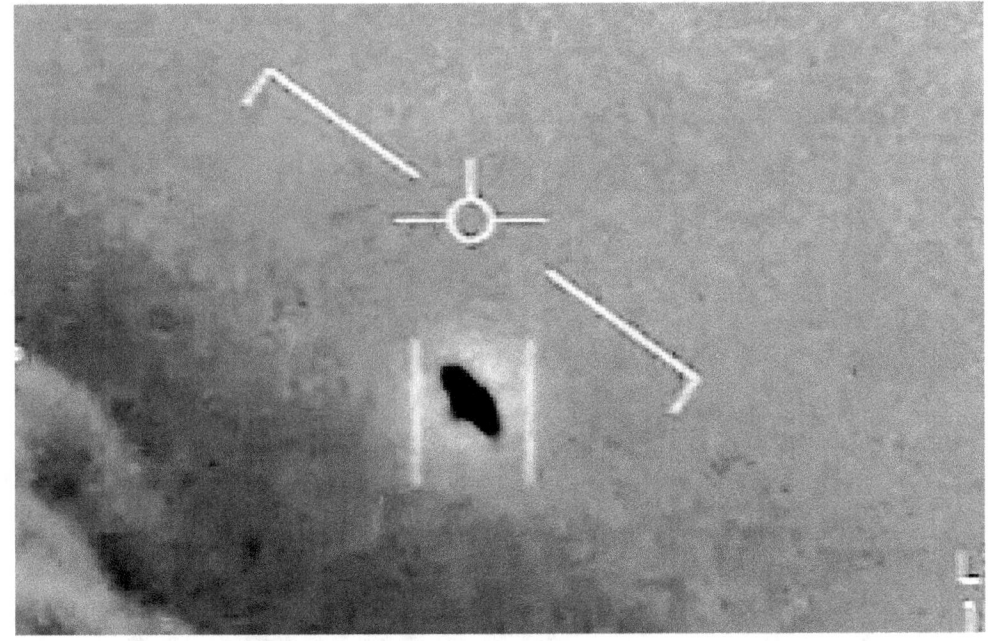

Footage of bizarre metallic UFO shown by Pentagon officials

Beyond photos are lots of sightings by people who are not easily fooled by things like reflections, smoke, and mirrors. But again, sadly, there are a few unscrupulous Earth people who manage to create uncertainty by various fakes.

There are plenty of rumors circulating (surely, you have heard some) that claim the U.S.A. government has secretly retrieved flying saucers, etc. and possibly extraterrestrial bodies (dead or alive ?). Have live extraterrestrials helped us with our extremely steep technological learning curve ? Or did they reverse-engineer some crashed technology ?

What do they look like ?

Many artists have created various images from their own imaginations. Usually, they seem to be influenced by horror movies. Nice artwork, but not very probable. Would an intelligent species travel here in order to kill the human species ? Not likely, because it is not logical nor useful to them. So, much more likely, images of extraterrestrials would be non-menacing in appearance.

Lots of people claim to have been abducted. This author is one of them. It is odd that so many images of extraterrestrials have been drawn which have a strikingly similar appearance. Not proof; simply odd ! See the following:

Typical reported appearance.

Our own species evolved from small creatures that walked on four appendages. Being small, we needed to be smarter than the big beasts in order to survive. Evolution is the survival of the fittest. Larger beasts did not need to be brighter, just more powerful and much quicker.

We walk upright on two legs because we needed to see farther so that predators could be seen at a distance. Also, we badly needed hands with fingers and opposable thumbs to use tools. Why ? Because we are not huge, powerful, fast, carnivore predators; we are small, weak omnivores needing to find vegetable food sources, and to catch and kill smaller meat/fish sources. Most likely, extraterrestrials would also be upright with hands and opposable thumbs.

Our binary vision (two eyes looking forward) gives us three-dimensional views so that we can determine distances. Most likely, extraterrestrials would evolve similarly with forward-looking three-dimensional vision, although more than two eyes is a distinct possibility.

The nerve bundles that carry eye data to the computer (brain) for processing are large, fragile, and need strong protection. Most likely, a brain-case, such as a skull, would be the best way to evolve this critical function. A 'head' of sorts could be expected, because a sphere-like shape contains more volume with less surface area. It is not surprising that the crucial first four senses, (sight, hearing, smell, and taste) are in our protective heads.

Noses ? Probably. A sense of smell might help to warn of an enemy nearby, or food that is too rotten to eat. There may be other good reasons for a sense of smell, such as pheromones. Sex ? Necessary for genus survival.

Eat ? Of course; energy must come from something and enter a body somewhere, somehow. A mouth ?

However, that does not preclude other means of energy entering an entity. Plants do it with solar energy captured by leaves and requiring CO_2 and H_2O to produce sugars (energy) and O_2 (oxygen). A photosynthetic extraterrestrial is possible, but quite unlikely due to the smaller amount of energy available in light. A very large leaf-like surface area would be required. Big surfaces might be heavier or more bulky; unlikely.

Burning food for energy also requires a means of inputting an oxidizer and outputting a gas (some form of smoke from burning energy). A breathing orifice ? Excreting the ash is also necessary for food eaters. What ? Oh yeah, that too...

Energy distribution to the muscles and organs throughout a body seems to be a logical necessity. The implication follows that a pump (heart ?) and energy-carrying liquid (blood ?) of some sort could be expected. Tubes to carry the liquid (arteries and veins ?) could also be expected.

Binary ears are also a necessity for three-dimensional hearing in order to sense where sounds come from. Some form of sound-collection appendages should be expected. Perhaps even more than two ear-like appendages or orifices might have evolved. Ears made us Earthlings safer and provided good communication. That brings up another point: groups.

Being relatively small and vulnerable, early evolutionary humans needed numbers to cooperate in hunting, protection, and community work. Reproduction is more successful for smaller beings in larger groups to prevent inbreeding DNA mistakes.

Therefore, upright walking beings with skulls (heads) and appendages with opposable thumbs, at least two eyes and ears seem most likely. Furthermore, they would probably have a nose of some sort and a mouth of some type. If they look a bit like us, it is not surprising.

Yup, me too...
...or four !

What do they eat ?

If they eat ! Probably. Considering the above, (**What do they look like ?**), the energy intake of extraterrestrials would probably involve food of some sort, much like us. They are possibly carnivorous, but unlikely solar powered plant-like or herbivorous beings due to the smaller amount of energy per mass in plants. Most probably, extraterrestrials are omnivorous just as we humans are. For survival of the fittest, evolution would choose those with healthier, more powerful diets. That does not mean that they are here to eat us !

What do they smell like ?

A guess would be: probably not too strongly stenchy; otherwise it would have made them more vulnerable in their own evolutionary process. All Earth animals have smells due to the processes of life. Converting food energy to muscle energy produces various chemicals, some of which are quite pungent and gassy. There seems to be no reason to suggest that extraterrestrials would be much different. We may like their smell-- or not ! And of course, vice versa in spades !!

What senses do they have ?

Again, considering the above, (**What do they look like ?**), almost certainly an extraterrestrial would need all of the senses that we have (touch, taste, smell, sight, and hearing). Perhaps they may have another one or more than our five. Soundless mental communication (telepathy ?) may have evolved, i.e. ESP. Nothing precludes it; it might have been necessary for survival ? Many abductees have reported soundless voices in their heads, and in their own languages and accents.

What might they like ?

What does any intelligent being enjoy ? **Security ! Life !** Second, lets consider personal likes. The state of being satiated or glutted; fullness of gratification, either of the appetite or of any sensual desire. Maybe comfort such as temperature, softness, good tasting food, and/or happiness are probable.

A feeling of security belongs in the category of community (groups). Camaraderie, friendship, loyalty, togetherness, teamwork, even love is quite likely...

Sex ? Hmmmm... Necessary for survival, so... Why not ? With us ? Not likely; probably too much DNA differences for successful reproduction. However, with such advanced technology as they must have, some chromosomal modifications could make that happen ! Fun ? ? ?

What diseases might plague them ?

Disease means the lack of ease. Technically, a broken arm is a disease. A headache is a disease. But a disease caused by an external and/or contagious something (germ, virus, chemical, parasite, etc.) depends upon the environment where the affected being exists. For an unknown being from an unknown environment, that is a very difficult question to speculate about. It is also extremely important for contact between races !

Diseases are not only abundant but also rampantly varied in every species of animals and plants on our Earth. They also change in nature as time passes. Surely, extraterrestrials have diseases too, unless they have discovered some medical 'miracle', but those diseases are probably not our diseases. On the other hand, some may be transferrable to us and vice/versa, especially bacteria and parasites.

Some rumors exist that alien technology has been reverse engineered in the pharmaceutical field as well, and those drugs are currently being distributed to the population at large in order to pre-immunize us from their diseases-- thusly allowing imminent contact ! **Oh my !!**

What do they want from us, if anything ?

Could it be camaraderie, friendship ? Trade, technology exchange, variety of goods, cultural exchange, simple scientific curiosity... all are probable. Perhaps they have need of something we have in abundance and their planet has little. Do we need anything from them ? That could be so.

Maybe their DNA **is** similar enough to interbreed. Perhaps that possibility would require some gene modification and engineering. It will be interesting to find out when open contact becomes common. And open contact **will** eventually be common. No, they are not searching for meat !

What help may they offer to us Earthlings ?

At this writing, we apparently are not technically advanced enough to travel to another planet; they are ! So, the answer is technology, of course. Technology of any and all sorts. Better and truer philosophy ? How to live peacefully ? Oh, we should definitely hope so !

What should we worry about, if anything ?

A person's cultural identity, or their self-conception and self-perception is related to nationality, religion, ethnicity, social class, even locality. We should worry that many cultures on this old Earth are quite resistant to differences and to unusual beliefs, and even unusual or surprising truths. Religion will be a most interesting but volatile topic.

Almost all organized religions will require major revisions. Some of those cultural groups (or individuals) will surely reject or even hate extraterrestrial contact. No doubt, there will be some trouble. Probably a bunch ! What if the extraterrestrials are not religious ? What if they are ? What if they believed in a single God ? What if it is the same God as the one who created our universe ?

Some single-God-based religions here on Earth might have a great reluctance to share the same God with other worlds. Atheists probably would again deny the existence of any God. Perhaps some of them would become believers.

Hopefully not much trouble would occur ! Friendly acceptance and welcoming would certainly be our very best option, although, sadly, that is unlikely. Perhaps we Earthlings aren't smart enough yet to accept new realities. Perhaps the extraterrestrials will simply watch us being stupid forever. That would be a very sad state of affairs.

Actually, there is no reason whatsoever that belief systems must all agree or be the same. All that is truly required is: agree to disagree gently; live and let live. Why care what who believes as long as they are peaceful. Well... except for the laws of physics, which are unbreakable as near as we know.

Perhaps we should enthusiastically concentrate on making some global **changes ? Yes !**

<u>**End of Chapter 1**</u>

Chapter two –   How ?

How big is our solar system ?   Page 11
How many stars are there ?   Page 16
How big is our galaxy ?   Page 17
How big and how old is our universe ?   Page 18
How many planets support life ?   Page 18
How many different types of ETs might there be ?   Page 19
How many ETs might be capable of traveling here ?   Page 19

In order to fully appreciate the questions and answers in this book, it is important to realize the immensity of the main topic-- *Are They Here ?* So let us start this chapter with orientation regarding sizes, quantities, and numbers. Some of the numbers are going to be huge, but do not despair; do not shy away from them. They are real, necessary, and easy to understand. The reader might already know most or all of this first part. Reviewing it again could be interesting. You might be surprised !

How big is our solar system ?

Big ! A great place to begin is our sun, Sol.

SOL
It is a medium sized yellow star. It is a huge ball of extremely hot gaseous plasma. Hydrogen is 90% of the gas, helium is about 9% and a few slightly heavier elements make up the rest (1%). Surely, you remember that hydrogen is the smallest and lightest element in the periodic table, and helium is the second lightest ? Yes, I remember, and quit calling me Shirley; my name is John.

The distance from the surface to the core center is 432,000 miles (diameter is 864,000 miles). To compare, Earth's radius is only 8000 miles and diameter is only 16,000 miles. Even though Sol is made of the two lightest gasses, due to its huge size and mass Sol's monstrous gravity is sufficient to hold all of our planets in their orbits.

At the core, gravitation pressure forces nuclear fusion to occur, fusing two hydrogen atoms together to produce helium. This process releases fusion energy just as Einstein's formula states: Energy = Mass * c * c where 'c' is the speed of light (c = 186,000 miles per second). 'c' times 'c' (or 'c' squared) is a very large number ! Therefore, a huge mass multiplied by 'c' squared equates to a monstrous amount of energy being produces constantly. In other words, it is basically an unbelievably large, continuous H-bomb !

Sol is speeding in its orbit around our galactic center at about 536,865 mph. Plus, our galaxy is also traveling toward the Hydra constellation at about 1.34 million mph. The speeds combine of course.

The temperature of the sun varies from around 27 million degrees F at the core to only about 10,000 degrees F at the surface, according to NASA. No life here !

* * *

Mercury
The nearest planet to Sol is Mercury. its diameter is 3,000 miles and distance from Sol is 36 million miles. Surface temperature in sunshine is 800 degrees F, and night temperature is 290 degrees F below zero, due to having no atmosphere. No moons here. No life here.

* * *

Venus
The next planet outward is Venus at about 67 million miles. its diameter is 7,500 miles, just a bit smaller than Earth, but much hotter. Orbital distance (meaning closest possible when our orbital positions are lined up) from us is about 26 million miles. Its dense atmosphere holds heat in. Surface temperature is about 870 degrees F, although about 30 miles up, atmospheric temperature ranges are similar to Earth. That doesn't mean it is livable, the air is 97% CO_2. No moons here. No life likely.

* * *

Earth
Next, us. Earth. We are 93 million miles out and diameter is 16,000 miles. We travel in our solar orbit 584,336,233 miles (yup, millions) each year, which is 1,599,825 miles each day at an average speed of 66,659 mph, whilst we spin at about 1000 mph at the equator (zero at the North and South poles). Complex motions when we add in the motions of Sol and our galaxy ! It is good that we don't feel it, due to constant speeds and good old Earth gravity holding us down quite snugly. Also, the sky is moving at the same speeds in the same complicated manner.

You know the rest. One moon and gobs of life in a plethora of beautiful environments, forms, types, and places.

* * *

Mars
Our next neighbor is Mars. It is a cold, dusty, desert world with a very thin atmosphere. Diameter is 4,200 miles (about half our size) and Solar distance is 142 million miles. Orbital distance from us is about 49 million miles. its atmosphere is over 100 times thinner than Earth's and mostly CO_2,

nitrogen, and argon. Temperature range is as high as 70 degrees F and as low as about 225 degrees F below zero. It has some water but very scarce on the surface. Two small moons. Possible life, but with little oxygen, maybe only plants. They love CO2 !

***

Asteroid Belt
This is a huge, wide randomly mixed belt of particles and chunks. The main belt lies between Mars and Jupiter, roughly two to four times the Sol to Earth distance, and is a region about 140 million miles wide. No piece of rock in it is very large and the total mass of the entire belt is smaller than our moon, Luna. The belt orbits Sol and is similar to the rings of Saturn in appearance.

* * *

Jupiter
The next planet outward is Jupiter, a huge gas giant. It is almost a proto-star due to its size, twice as massive as all the other planets combined. Diameter is 48,000 miles and solar distance is 484 million miles. Orbital distance from us is about 391 million miles. Jupiter has 95 moons that are officially recognized by the International Astronomical Union. The four largest moons are Io, Europa, Ganymede, and Callisto.

The following facts are copied from NASA: Io is the most volcanically active body in the solar system. Ganymede is the largest moon in the entire solar system (even bigger than the planet Mercury). Callisto's very few small craters indicate a small degree of current surface activity. A liquid-water ocean with the ingredients for life may lie beneath the frozen crust of Europa, the target of NASA's Europa Clipper mission slated to launch in 2024.

Jupiter's environment is probably not conducive to life as we know it. The temperatures, pressures, and materials that characterize this planet are most likely too extreme and volatile for organisms to adapt to.

***

Saturn
The next planet outward is Saturn, one of the most beautiful planets in the solar system with its colored bands and gorgeous rings.   Like fellow gas giant Jupiter, Saturn is a massive gas ball made mostly of hydrogen and helium. Diameter is about 75,000 miles and solar distance is 886 million miles. Orbital distance from us is about 793 million miles.

Saturn is not the only planet to have rings, but none are as spectacular or as complex as Saturn's. Saturn also has dozens of moons.

From the jets of water that spray from Saturn's moon Enceladus to the

Page 13

methane lakes on smoggy Titan, the Saturn system is a rich source of scientific discovery and still holds many mysteries.

Saturn's environment is not conducive to life as we know it. The temperatures, pressures, and materials that characterize this planet are most likely too extreme and volatile for organisms to adapt to.

While planet Saturn is an unlikely place for living things to take hold, the same is not true of some of its many moons. Satellites like Enceladus and Titan, home to internal oceans, could possibly support life.

A montage of Saturn and some moons courtesy of NASA

* * *

Uranus

The next planet outward is Uranus. Yes, the pronunciation is as close as you can get to "your anus" ! Uranus is the seventh planet from the Sun, and it's the third largest planet in our solar system – about four times wider than Earth – about 64,000 miles. Distance from Sol is about 1,780,100,000 miles (Yup, 1.8 billion). Orbital distance from us is about 1.687 billion miles. It takes light two hours and 43 minutes to travel from Sol. A round trip radio communication from Earth takes about 5 hours and 10 minutes. Uranus takes about 17 hours to rotate once (a Uranian day), and about 84 Earth years to orbit Sol (a Uranian year).

Uranus is a very cold and windy planet. It is surrounded by faint rings, and more than two dozen small moons as it rotates at a nearly 90-degree angle from the plane of its orbit. This unique tilt makes Uranus appear to spin on its side. Uranus is blue-green in color due to large amounts of methane, which absorbs red light but allows blues to be reflected back into space.

Uranus is an ice giant. Most of its mass is a hot, dense fluid of "icy" materials – water, methane and ammonia – above a small rocky core. It has an atmosphere made mostly of molecular hydrogen and atomic helium, with a small amount of methane. Weird to say the least. It has 27 known moons and rings similar to Saturn's. Voyager 2 is the only spacecraft to fly by Uranus. No spacecraft has orbited this distant planet to study it at length and up close. Life is improbable.

***

Neptune

The last true planet outward is Neptune, number 8. Its diameter is about 30,775 miles and solar distance is about 2.8 billion miles. Round trip radio time for communication is over 8 hours.

Dark, cold, and whipped by supersonic winds, ice giant Neptune is more than 30 times as far from the Sun as Earth. Neptune is the only planet in our solar system not visible to the naked eye. In 2011 Neptune completed its first 165-year orbit since its discovery in 1846.

Neptune is so far from the Sun that high noon on the big blue planet would seem like dim twilight to us. The warm light we see here on our home planet is roughly 900 times as bright as sunlight on Neptune. its diameter is 30,775 miles. Solar distance is about 2.8 billion miles.

Neptune is similar to the other two gas giants but, scientists think there might be an ocean of super hot water under Neptune's cold clouds. It does not boil away because incredibly high pressure keeps it locked inside.

Neptune has 16 known moons. Neptune's largest moon is Triton. It is extremely cold, with surface temperatures around minus 391 degrees F. And yet, despite this deep freeze at Triton, Voyager 2 discovered geysers spewing icy material upward more than 5 miles. Triton's thin atmosphere, also discovered by Voyager, has been detected from Earth several times since, and is growing warmer, but scientists do not yet know why.

Life on Neptune or its big moon Triton is highly unlikely.

* * *

Pluto

Poor little Pluto was long considered our ninth planet, but the

International Astronomical Union reclassified Pluto as a dwarf planet in 2006. Pluto is about 1,400 miles wide. That's about half the width of the United States, or 2/3 the width of Earth's moon. Its elliptical orbit (which oddly is often inside Neptune's orbit) averages about 3.6 billion miles from Sol. It is too small and far too cold to support life.

***

Why so much astronomy ? Logic and facts are used to answer questions, especially those questions that require speculation. When facts are more accurate the guess becomes closer to true. Since possibilities of life and of long-distance travel are questioned, numbers are required for answers.

Consider this fact: With a fast rocket, it takes us around three days to reach the moon; approximately seven months to get the closest planet, Mars; 15 months to reach Venus; six years to reach Jupiter; seven to reach Saturn; 8.5 years to reach Uranus; 9.5 years to reach Pluto (currently); and twelve years to get to Neptune the farthest planet.

And so, the above constitutes our Solar System, our chunks of dirt around our star, Sol. And that concludes the review of very nearby space. The rest is cold, empty vacuum for a **very, very, extremely long distance !**

How many stars are there ?

The night Sky courtesy of Hans at PIXABAY

When we look up on a clear night without city lights, we see a bunch of dots of light. We usually think of them as stars. Only a few of what we see are actually stars. A few of those points of light are our own planets and the rest are mostly galaxies much like ours, not single stars. Those galaxies are so far away that they seem to be a single point, but that is definitely not the case. Furthermore, galaxies contain lots of stars. Sol is a star in our own galaxy, 'The Milky Way'.

How big is our galaxy ?

Really Big ! In order to speak or write about the size of our Galaxy or our Universe, monstrous numbers are required, and there would be way too many zeros at the end of numbers to compare or even to comprehend.

Scientist use a very neat method of measuring distances, but it requires a tiny twist in our thinking. It is called a light-year. Remember the Einstein speed limit of the Universe: It is 'c', the speed of light, which is 186,282 miles per second, or in the metric system, about 300 million meters per second. The distance that a photon, moving at its normal speed, 'c', will travel in a year is called a light-year. So it is not a time; it is a distance.

Just for fun, it is miles per second times seconds per minute times minutes per hour times hours per day times days per year: So-- 186,282 * 60 * 60 * 24 * 365.25 = 5,878,612,843,200 miles. That's almost 6 trillion miles !

Now we have a measuring stick that is long enough to measure galaxy sizes and the Universe. This measuring stick is also very convenient to use for determining a travel time. If we know the distance (in light-years) to the nearest star, Proxima Centauri, we know how long it will take to get there (*if we were a photon of light*). Proxima Centauri is the closest star to Sol at 4.2465 light-years. Just divide that number by our fractional speed compared to 'c'.

Example: A spacecraft traveling at one tenth of one percent of the speed of light, would take more than 4 divided by 0.001 years (that is 4 thousand years !) and that slow speed is only 670,615,200 mph (yup-- 670+ million mph for four thousand years). Not with our space rockets !

Our galaxy, 'The Milky Way' is a spiral-armed disk. It is thicker at the center than at the edge. There is a massive black hole at the center of our galaxy. Black holes are super-massive compacted stars with so much gravity that nothing can escape from it, not even light !

Distance across our Milky Way is about 1.9 million light-years (if we include the recently discovered dark matter extension). So it's nearly 12 quintillion miles wide. **It's pretty big !**

The nearest galaxy to ours is 'Andromeda Galaxy', which is 2.5 million light-years away from Earth. It is quite wide, too ! They all are.

Andromeda – Our Nearest Neighbor Galaxy courtesy Space.com

How big and how old is our universe ?

Huge ! And Very !! Our Universe is speculated to be about 93 billion light-years (or more) in diameter. And it is expanding. Why bother with calculating the distance in miles ? **Too big !**

Oh, alright, we will calculate it: It's about 564+ with 21 zeros.

Age ? Our universe is approximately 13.8 billion years old but its exact age is not yet clear. What we do know is that it's likely less than 14 billion years old. If controversial measurements of the expansion rate of our universe are correct, the cosmos could be a little younger. The uncertainty is not because our methods of measuring our universe's age are bad. Rather, there are still things about our universe we don't understand. Yet !

How many planets support life ?

Lots ! Logically, that has to be true, simply because of the number of planets that exist in the our huge Universe.

While estimates among different experts vary, an acceptable range is between 100 billion and 2 trillion galaxies exist in the observable Universe.

There are about 100-400 billion stars in an average galaxy such as ours. Some of those stars have planets. Probably, most stars have several planets !

There are estimated to be at least 100 billion (maybe 200 billion) planets, just in our galaxy, 'The Milky Way'.

Data from NASA: Scientists have estimated that 1 in 5 stars like our Sol has at least one Earth-like planet orbiting around them, which may support life. That is because of the 'Goldilocks' filter idea: not too big (too much gravity); not too small either, not too close to its star (too hot); not too far from its star (too cold); but just right !

Based upon the mapping of our Milky Way, and through simulations, there are an estimated 40 billion planets (possibly more) that might support life, just in our own Milky Way galaxy.

Therefore, a good estimate of planets in our Universe that could support life would be 100 billion (using the lower estimate of the number of galaxies) times the above. Very conservative answer:

at least *4,000,000,000,000,000,000,000*

That is 4 sextillion life-possible planets similar to Earth.

This huge number of planets does not consider the probability that, like our own Jupiter and Saturn, sometimes moons of such larger planets might have conditions that could support life. At least, we should double the number above to include those such moons-- so 80 quintillion 'Goldilocks space-balls' !

How many different types of extraterrestrials might there be ?

That question is quite easy and, at the same time, extremely difficult. Lets limit the question to 'intelligent extraterrestrials'. That should bring the number down a long way. With that 40 to 80 quintillion number just above, one could say perhaps one percent of that huge number of life-able planets could produce **intelligent** beings. It is simply a wild guess but possible. That still leaves about *400 to 800 quadrillion different intelligent races in our Universe.*

How many extraterrestrials might be capable of traveling here ?

Inter-galactic travel seems, at this time, not very probable due to the extreme distances involved. Compare the distance between stars in a galaxy with the distance between galaxies. The nearest star is a bit over 4 light years; the nearest galaxy is 2.5 million light-years. It is difficult to imagine that kind of travel, so let us eliminate all other galaxies as candidates for extraterrestrials who might get here.

So lets divide our previous number (400 to 800 quadrillion) by 100 billion galaxies. *Then the possible number of intelligence races in our galaxy drops to only 4 or 8 million* if we include some moons. That might still be a very conservative number. However, it doesn't answer the question of capability.

In order to consider capability, we must consider how long it took for life to begin on our planet, then evolve into an intelligent species, then to develop science to the point where that species could achieve interstellar travel (that is, star to star).

Consider that, in our 13 or 14 billion year old Universe, each one of its stars has a different birthdate. Those star-birthdates could be separated by millions of years.

Each planet everywhere has its own different birthdate. Those dates vary, not by days, but by hundreds, thousand, millions, or maybe billions of years.

That fact implies that some intelligent species are far behind us, and some are far ahead of us (**and NOT by just a few hundred years !**)

If civilizations occurred on some of those planets, when did their technology reach a capability for space travel ? Some sooner than others, obviously. What will our own technology be capable of in just another piddling hundred years ? How about a thousand ? Surely, technology in all of the civilizations that exist differs by a few or else by many years, even by lits of thousands of years ! No doubt **those extraterrestrials can travel through space !**

If only half are advanced enough to get here, that leaves **only 2 to 4 million in our galaxy.** Considering the monstrous distance across our galaxy, the farthest ones, perhaps, should be eliminated as probabilities. Maybe not, but... Even with technology that does not require slow space ships (warp speeds ?), some distances might be too far to be possible.

Our galactic disc has spiral arms, and we are on one of them about 2/3 out from the center. If we eliminate stars (with planets) that are **not** in our neighborhood, suppose 95% of them, then the number of probable space travelers drops to 5% of 2 to 4 million, leaving about 100 to 200 thousand. That could easily be wrong, but most surely it is not zero.

So, in summary: How many extraterrestrials could probably be capable of traveling to Earth ? About 100 to 200 thousand or more !

End of Chapter 2

Chapter three   –   More Hows ?

How could ETs travel such a distance to get here ?   Page 21
How difficult is the travel trip ?   Page 23
How might the trip difficulties be eased ?   Page 24
How did they get here, then ?   Page 29
How can we travel to their home ?   Page 30
How would we learn to communicate with them ?   Page 30

Before we begin, there is a solid fact that we must wrap our minds around: **Energy** is the most important of all subjects having to do with life, with survival, with our Universe, and with travel around in it. Remember Einstein's formula: Energy = Mass * (c squared). What this also says is Mass = Energy / (c squared). Mass and energy are transformable back and forth. However, a very important fact is that neither energy nor mass can be created or destroyed, only changed in type. P. S. Heat is a form of energy. More on this later...

How could extraterrestrials travel such a distance to get here ?

Almost certainly, given our current understanding of the laws of physics, they won't use slow rockets. Probably not fast rockets, either. The problem is power. More correctly and precisely, energy is the difficult part of travel. P.S. Energy is power multiplied by time.

The Space Shuttle

See the following website:

https://www.space.com/5094-alpha-centauri.html

Once in space, a vehicle has almost a perfect zero resistance because space is a vacuum. An object moving in a direction will continue moving in that direction with the same speed unless an outside force changes its speed or direction.

Achieving that speed requires acceleration, and acceleration requires energy. Solar energy won't be available once the space ship leaves its star system. Therefore, energy sources must be carried with the vehicle. That means mass; mass require more energy to accelerate the extra mass. This is why our space rockets are so huge in order to lift relatively small payloads.

Just for fun, lets assume a spaceship has plenty of fuel and we break free of Earth's gravitational pull. We aim for Proxima Centauri which is right next to Alpha Centauri, only about four light-years away. Suppose we can accelerate until we get *very near* to 'c', the universal speed limit; light-speed ! Warp Factor One !

That requires a monstrous amount of energy, and there are problems with the scenario:

The nearer we get to 'c', increasingly more energy is required to accelerate. To actually achieve light-speed, an infinite amount of energy is needed. Fuel ? Below is a graph showing energy requirements (upward) to try to achieve light-speed (increasing speed to the right). As you can see, the speed of light is only approachable asymptotically but not completely reachable (laws of physics):

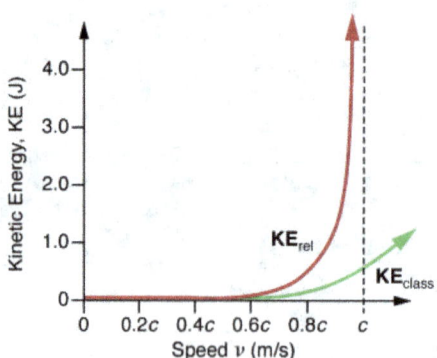

Clearly, any speed past about ½ of 'c' becomes way too tough.   (2)

Once we get to half light-speed, we could coast for part of the 4 years, but then we need to slow down. Deceleration requires exactly as much energy as it takes to accelerate this spaceship. So the final part of the journey is spent decelerating. The speed up and slow down parts take a lot of time, so the nearly four year trip will be extended. Time to Proxima Centauri 4.25 light-years away ?

A space ship might accelerate at 1 G for about a year to get to about ½ 'c' (half of light-speed). It would have traveled at an average speed of ¼ 'c'. Distance traveled would, of course, be about ¼ of a light-year. We must reserve time for the exact reverse procedure to decelerate at the end. In the middle time, we coast at ½ 'c' for 3 years and about 9 months. Total time would be about 5 years and 9 months.

It would be do-able if we had the energy source required for the acceleration/deceleration part. That rocket would require a total burn time of two years ! That is a load of fuel !! And that much more mass would require far more energy to accelerate it, plus the rocket's mass.

Using current rocket speeds, even with some fantastical propulsion methods and fuel, the time of travel would be long. One estimate (not yet do-able by far) was about 70+ years, unless we can accelerate slowly the entire time. We need a crew of smart 'Sleeping Beauty' astronauts. Then to get them back !

In summary, extraterrestrials probably do not use our type rockets for the entire trip. We will visit another method later. Meantime...

How difficult is the travel trip ?

Zero gravity is not an easy environment for humans to live with. It is a constant state of free fall, such as the first 15 seconds of parachuting from a plane. It feels like your falling off the Empire State Building. That takes a while to get partially used to.

Water must be carefully controlled and contained because if it gets loose, it floats around as a quivering ball. Any touch might send it into several smaller balls flying in several directions. Furthermore, ***all liquids*** must be recycled, since there are no water sources in space.

The necessary human eliminations of liquids and 'solids' are a constant and extremely difficult battle. Cleanliness is important and is also very difficult. A nice, hot shower won't be available. Without gravity, nothing falls; it just floats. Everything must be captured and recycled or ejected.

On a long trip, such as many months or years, even the 'solids' must be recycled, not ejected. Maybe a garden would be necessary. Sterilize ejectable 'solids' and use for fertilizer. Packaged food has to be stored and re-heated. One cannot chop up a fresh salad, or boil some bean soup in a pot because the liquid would simply float out of the pot.

Sleeping is not easy whilst feeling like you're falling. Some sort of tethering must be used to keep a person from floating out of bed.

Exercise is a necessity because muscles atrophy, bones get brittle, and organs begin to fail in zero-G. Even the process of exercise is difficult because of weightlessness. Astronauts use rubber strips to pull them toward the floor to exercise. Not quite the same thing as gravity weight.

Perhaps boredom could also be a problem. Living areas cannot be very large, so astronauts can't be claustrophobic. Mental health is extremely important since there is no escape from their environment.

There is always a tiny chance that a small chunk of rock, moving at an incredible speed, might punch a hole in the hull. Any equipment failure could become instantly deadly. There are no rescue squads, firetrucks, or ambulances. Sickness could be fatal. Sounds like fun ! Astronauts are true heroes; very brave heroes !

How might the trip difficulties be eased ?

Energy problems:
Nuclear power for electricity would not require as much mass as other means, and electricity is not just a luxury-- it is a necessity. Laser fusion is (finally, recently) getting off the ground at the Lawrence Livermore National Laboratory, but it will be a while before it is up-scaled and generally usable. Battery power is out of the question as a main source. Other energy producers might arrive.

Food problems:
A large garden could make a big difference for voyages that last more than a few months. A large amount of room would be required, though. Water is once again a problem, not to mention humidity/condensation/recycling for drinking.

Energy for lighting will cost the system; much is wasted for lighting up plants. And then there is fertilizer recycling... Yup, just imagine... However, food storage is easy; baby it's cold outside !

Weightless problems:
Artificial gravity ! Having 'weight' would solve many of the problems listed above, but not all. The idea is simple enough-- create a living space shaped like a hamster exercise cage. It would have to be huge, and it would spin slowly. Centripetal force would create artificial 'weight' outward on its outer surface. Then one could walk with feet pointed outward, head pointing to the center, and feel quite normal. Toilets would work ! It would take a small amount of energy to run it due to bearing friction. A beautiful example was in a 1968 movie called: "2001: A Space Odyssey", a science fiction film produced and directed by Stanley Kubrick. The graphics and music in that movie are truly fantastic. Here below is an image from the movie:

A rotating wheel to produce artificial gravity in space

Heating and cooling problems:

Once a space vehicle is no longer near a star (Sol or ?), temperature control is a tough problem. Without external radiation coming in, heat energy will radiate out from an object in space. There is no heat conduction in a vacuum.

A vacuum is a space where there is nothing, Nothing exists in a vacuum. Nothing cannot have a temperature because (almost) nothing is there. Therefore, the temperature of outer space does not exist in the way we ordinarily think of temperature. However, **things** in space can have a temperature.

Just remember that 'cold' does not exist (Sort of like a vacuum, yes ?). Heat *does* exist. Heat is one form of energy. Coldness happens when there is a lack of heat. When there is no heat at all, then a condition occurs known as 'absolute zero temperature', when the thermometer reads: 0° Kelvin, -273.15 ° Centigrade, or -459.67 ° Fahrenheit. Nothing can get colder than that.

So, an object in outer space with no significant star incoming radiation (such as Sol) would soon radiate away all of its own heat until its temperature would become (or at least approach) absolute zero. Really superb insulation would slow that process greatly, but ultimately, the object would lose all of its heat. This is a problem; see below: drinking water problems.

Conversely, an object inside the solar system would absorb radiant energy from our star and heat up rapidly. Again, superb insulation and reflective coverings would slow that process greatly, but ultimately, the object would become extremely hot on the 'sunshine' side.

The International Space Station handles the problem with a complicated system of heat exchangers and some superb insulation that wraps around the entire station except for windows and solar panels. The station rotates slowly relative to Sol. Temperature varies rapidly from +250 ° F in sunlight to -250 ° F in shade. It is a constant battle.

Copied from space.com:

The International Space Station is the largest structure ever built in space. The first module-- the Russian Zarya module-- was launched from Earth on Nov. 20 1998. International crews have continuously occupied the orbiting space laboratory since 2000, and five space agencies contributed to the building of the structure.

The $100 billion ISS has the wingspan of a football field and the living space of a five-bedroom house. It took more than 115 spaceflights of different kinds of spacecraft to build the station, and astronauts and cosmonauts have spent more than 1,000 hours on spacewalks outside of the station.

The solar panels are plentiful and huge due to the large energy requirements of this huge orbiting satellite. A picture is below:

The International Space Station
Image courtesy of NASA

Drinking water problems:

Even though recycling *everything* (almost) is necessary, some loss is bound to occur. Carrying lots of water is necessary and massive (the term heavy in gravity equate to massive in zero-G). Tanks of oxygen and hydrogen could be carried to be combined making water, but the mass is the same. However, the carefully controlled combustion of the two gases ($2H_2 + O_2 = 2H_2O$ + energy) will produce pure water and a lot of heat. This heat energy could be a source to replace heat loss in dark space. Also, the violent reaction of combustion could be an energy source for small maneuvers of the space ship.

Even though we speak of space as 'empty', there are bits and pieces of dust and rocks and some atoms floating around. These are extremely sparse in most places out in deep space. Nevertheless, hydrogen atoms, left over from the 'Big Bang' of ~13.7 billion years ago are sparsely 'plentiful'.

At the high speeds of a space ship, over time, hydrogen atoms could, perhaps, be scooped in to the front of the vehicle. This could help reduce the size of the initial load of hydrogen. Oxygen atoms out there are probably too sparse to scoop. This is a highly speculative idea, but might be possible.

Acceleration/Deceleration Problems

One of the laws of motion is (yup, physics, now don't get scared; it's simple) Newton's first law: "A piece of mass remains at rest, or in motion at a constant speed in a straight line, unless a force acts upon it." Example:

You press the gas pedal; your car's engine applies some force to the wheels and it starts to move. You press harder; it increases its speed quicker. If you do nothing more, it will coast (for a while-- friction, you know). You hit the brake pedal; it slows down. You crank hard on the steering wheel and your car changes direction. But you knew all of that. :)

In empty space, there is no friction to slow a piece of mass. Once a space ship is in space and headed out for a trip, it must be accelerated (speed increased) by a force. The amount of force applied to its rear determines the acceleration. The eventual speed depends on the length of time the acceleration is applied. Even a small force will move a space ship, but acceleration will be tiny. However, if that small force remains constant, the ship will continue gaining speed (accelerating) at a constant rate.

If the applied force is huge, the acceleration will be quick, however the occupants of the space ship will be subjected to a 'weight' relative to the rear of the ship. Sometimes, this is called the G-force.

Speaking of forces, consider NHRA's famous drag racer, John Force.

John Force's Funny Car

With a 'Funny Car' being forced to start down the 1000 yard track by a V8 engine producing nearly 13,000 horsepower, reaching a speed of 330+ mph in 3.4 seconds at the finish line-- John Force is mashed back into his seat with a G-force of almost 5. That means he feels like he weighs around 1,000 pounds relative to the rear of his car and still 1 G-force downward.

We cannot stand such 'G-forces' for an extended time, so our acceleration force must be reasonable. That's OK; even a gentle, but constant, acceleration will eventually produce great speed. Then, there is the next problem: stopping at our destination. This requires the same acceleration in reverse (deceleration). This was mentioned earlier...

So, halfway through the journey, we must switch the force from the rear to the front of our space ship. That means that half of our journey time is spent slowing down. That takes energy. John Force's Funny Car uses a couple of parachutes to help the brakes slow him down.

There is one interesting point to this acceleration/deceleration scheme, and it could be very important: during the first half (That is assuming there is no coasting middle part). If we choose a one-G acceleration, toilets would work if our 'floor' was oriented to the rear of the space ship. Then, during the second half, if we choose a one-G deceleration, toilets would still work if we rotate the space ship to point its rear forward. That is a superb idea because we won't have to add engines to the front of our space ship.

Overcoming almost all of the bad health side-effects of zero-G is extremely important. However if there is a long coasting time in the middle of the journey (years ?), then a large 'hamster wheel' as in that movie mentioned previously would be necessary.

In summary, extremely long space journeys are nearly impossible with our current technology, and might require more than one lifetime. There may be other methods, other technologies, but we still must obey the laws of physics.

How did they get here, then ?

Not with our known do-able technology ! But **Worm Holes !** Probably ? Maybe. A simple explanation is that a worm hole through space is similar to a railroad tunnel through a mountain. Einstein predicted them in theory. Basically, gravity warps space and it warps time (the fourth dimension) as well. If the warp-bend is strong enough, (think of a wrinkled blanket) then the distance between two far distant points of space is much shorter. The time through the worm hole would be far less.

Gravity warps the four dimensional space-time continuum due to mass. Remember that mass and energy are interchangeable. That implies that energy can also warp space-time.

In order to visualize warp, first look at just a two dimensional plane like a trampoline. Place a bowling ball in the center. Next, imagine a marble aimed from one side to the other, past the bowling ball. Its straight line motion will be bent around the bowling ball (or even captured by it). Then think in the third dimension, the height plane, which is also warped. Finally, imagine a fourth dimension, time, which is warped in the same manner. The implication is that **time travel just might be possible ! Wow !!**

To see a video of this light-bending thing, go to this website:

https://science.nasa.gov/universe/how-gravity-warps-light/

A very strange observation proved the space warping theory. An astronomer was watching a very distant star which was visible far, far away, behind Sol. As Earth continued in its orbit, the distant star's image should have disappeared at a known time, being obscured by Sol's image. It did not ! It remained visible for some time after it was behind Sol, due to that star's light traveling in a warped line as it went (in a straight line) past our star. Sol's massive gravity had actually 'bent' the straight path of light. Einstein was right ! As usual !!

For a more scientific discussion of the subject of worm holes:

https://www.astronomy.com/science/what-are-wormholes-an-astrophysicist-explains-these-shortcuts-through-space/

In order to travel through a worm hole, it must be stable, usable, and find-able. There seems to be just such a worm hole at Skinwalker Ranch. Again, it is worth a long look:

https://www.history.com/shows/the-secret-of-skinwalker-ranch

Apparently, that worm hole has been used by extraterrestrials for a long time, although we probably don't know how to use it (assuming it is still usable). If we do know how to use it, that must be an extremely well-kept secret by our government. Unlikely.

Evidence exists that there might be several other worm holes on our home planet, Earth. Hopefully, some day soon, we will all know. ***And we will meet The Extraterrestrials ! Maybe !***

How can we travel to their home ?

Same way ! Most probably, worm hole travel is the only way for living beings to achieve inter-stellar travel. Distances are too large for other methods.

We need outside help or else we must continue improving our understanding of our Universe and our technology based upon such learning. Outside help would be very welcome. That help would come from the extraterrestrials.

And that would depend upon our acceptance of them as well as their acceptance of us (a much more 'iffy' proposition !). We have religious and ethnic wars, homeless bums, street beggars, druggies, corrupt politicians, dictator governments, child sexual trafficking, and plenty of murders, thievery and other crimes. Would a successful, decent civilization touch us with a very long stick ? More on this later...

How would we learn to communicate with them ?

In order to learn/teach communication with other people who speak a different language or communicate in a different way, we need common ground of some sort. Peaceful gestures might be good. Offering a gift could help. And of course, these two beginnings go both ways. Then, comparing things and naming them back and forth was always the usual way, especially for primitive people. Are we the primitives ?

For technical cultures, numbers and physics work. Universal constants such as PI, the continuous ten-finger decimal series of 3.141592653... having to do with circles is a good beginning communication. What if they have 12 ?

End of Chapter 3

Chapter four –   Who ?

Who knows information about extraterrestrials ?   Page 31
Who should know about them ?   Page 32
Who would want them here ?   Page 33
Who would be in charge of welcoming them ?   Page 33
Who would not want them here ?   Page 34
Who are The Extraterrestrials ?   Page 34
Who are we ?   Page 35
Who wants to be the first to greet them ?   Page 36

Who knows information about extraterrestrials ?

The United States Government probably knows more than most other governments. It seems that, in 1947, a businessman in Washington State was flying in a small plane when he saw nine crescent-shaped objects flying past Mt. Rainier at a speed of (his estimate) several thousand mph. He reported that they were moving like saucers skipping across water. This was the beginning of the 'Flying Saucer' name for these Unidentified Flying Objects: UFOs. See more info here:

https://nuforc.org/

Also, in 1947, the 'Roswell Incident' occurred in New Mexico. A rancher on horseback found a 200 yard long debris field that looked very strange. He reported it and authorities investigated it. Soon, an Army Air Force officer announced to the press that it was a crashed flying saucer and they quickly cleaned up the debris field.

The following day, the Army refuted that announcement and said it was just a weather balloon. However, to anyone who had seen the debris (or the newspaper photographs of it, which are available), it was clear that whatever this thing was, it was no weather balloon. Many folks still believe that the crashed vehicle had not come from Earth. Multiple theories and stories have filled books every since. **Who knows ?**

https://www.history.com/topics/paranormal/roswell

There is a place north of Las Vegas, Nevada called Area 51. You can get to it on the UFO Highway, but you can't go in. It requires a security clearance that is a notch or two above Top Secret. Signs on the fence warn people that deadly force will be used on trespassers. Many people say that there are some extraterrestrials inside of a mountain there. Also, plenty of folks say that they have seen some extremely strange aerial oddities flying around there.

Our government has 'unclassified' UFO data at times, but has always, officially, denied facts and kept most information secret. Only in the last few years, they admit the possibility of UFOs but instead, wish to call them UAPs (Unidentified Aerial Phenomena). That's OK: the astronauts have seen UFOs in space for 60 years !

As for those who believe that these are simply U. S. government secret projects, some of them are-- but not all. Too advanced ! **Who knows ?**

Who should know about them ?

Shouldn't everyone know everything about them ? Maybe governments around the world keep denying UAPs and UFOs and extraterrestrials because they believe that knowledge would cause global panic. Remember the Orson Welles radio show ? It was based on H.G. Wells's 1898 novel, *'The War of the Worlds'.* The sci-fi book was about Martians landing in large numbers.

The night before Halloween 1938, Orson Welles and his radio show, *'Mercury Theatre on the Air'* performed a radio adaptation of H.G. Wells's novel about Martians landing with powerful weapons, and killing thousands of Earthlings. The radio drama was mostly 'news flashes' of how the Martians carried on their war against Earthlings. It was quite realistic. And yes, panic ensued across the land !

https://www.smithsonianmag.com/history/infamous-war-worlds-radio-broadcast-was-magnificent-fluke-180955180/

Maybe the Orson Welles incident, which had caused so many people to panic, has convinced governments to hold back information every since. But there are many people now who believe that *extraterrestrials are here already*, and even more people believe that they *do, in fact, exist*—and yet even more people believe that it is possible that they *might exist.*

None of the above mentioned 'believers' seem to be panicked. Of course they are not ! Logic tells us that, if extraterrestrials are technologically capable of traveling here, and are here, they would either: have no evil intentions, or they would already have cooked us all.

Every government of any kind should know all about the extraterrestrials. At some point in our future, a meeting of some sort will occur. Every one of our governments should be well prepared. We Earthlings all should be prepared and not scared.

However, some religions would be strongly opposed, and that needs to be addressed... the sooner, the better. Remember way back on April 12, 1633, the Catholic Church arrested the astronomer, Galileo, for his belief and his announcements that the earth revolves around the sun.

That interesting story is below:

https://www.history.com/this-day-in-history/galileo-is-accused-of-heresy

The looming question, then, is all about God. Naturally atheists would feel vindicated, claiming that there is no God. Especially if the extraterrestrials had no beliefs in God. But if the extraterrestrials did believe in a God (maybe they even have proof), other religions would wish to compare Gods, and probably argue vehemently. One big question would be: is there only one God, or one for each planet ? Hmmm... So, who should know about them ? **Everyone should know about them !**

Who would want them here ?

We should all want to get to know our neighbors, even if we deem them to be ugly. We should not be repulsed or frightened by their appearance. Imagine if there were no dogs, wolves, jackals, coyotes, dingos, hyenas... we had never seen any animal like dogs. And suppose a huge golden retriever walked out of a flying saucer on four legs, stood up, and with opposable thumbs, held up a sign with 3.141592653 (pi) written on it, and spoke, "Peace!" in a horrible, growling, sloppy, juicy, drooling accent. Would you think it was ugly or repulsive ? Watch a few episodes of 'Star Trek, The Next Generation' and begin to get used to odd looking beings.

Businesses of every sort should want to sell goods to them. Governments should all want to sign treaties with them. The entertainment industry would go wild with enthusiasm, in both directions. Philosophers would begin debate right away. Most of all, scientists would love the exchange of technologies. **We all want them here !**

Who would be in charge of welcoming them ?

Most likely, every government agency in the entire world would scratch and claw their way into a welcoming committee. It might be quite chaotic. Perhaps that is another reason that governments keep the entire subject secret and deny their existence.

Certainly, the United Nations would be a first choice, except for one big problem: the member countries cannot even agree among themselves. Also, many of those member governments are so corrupt and evil, that their participation would be quite embarrassing, interstellar-wise. Truth can be painful, yes ?

The best committee would include common citizens who would be fearless, fair, honest, non-political, open-minded, logical thinkers-- and well-educated to boot. It's a **BIG** committee ! There would have to be a few representatives from each nation. Those honest citizens would need the power to override the U. N. official governmental representatives. Do you know anyone like that ?

Naturally, interpreters would be required. That problem, alone, is formidable. Interpreters must be trustworthy and meet all of the above requirements of representatives as well.

Are we yet civilized enough to become a part of an inter-stellar community ? **Probably not ! So, who would be in charge of welcoming The Extraterrestrials ? We, the people of Earth !**

Who would not want them here ?

Many governments would not like smart, powerful, logical beings hanging around. Too embarrassing ! It seems quite logical that any civilization with the industry, technology, intelligence, and will to get here would also be extremely well governed. We could be. We should be. We would be-- except for evil, corruption, greed, poor education, and more...

Greed in this context does not mean that all Corporations are greedy. The best ones are simply desirous of making a good living, providing jobs, paying their investors, and providing goods and/or services to their customers. Without them, and their desire for a profit, we would have nothing.

So, who would not want them here ? Most certainly a few fanatics would not. And then there are **The bad guys !**

Who are The Extraterrestrials ?

Logically, they are certainly civilized people. Just different in appearance. Wouldn't it be interesting to have such neighbors living next door ? Imagine the exchange of dinners, the conversations about families, vacation places, politics, medicines, shoe bargains, tail-gate parties, recipes, hair styles, science, sports, babies (sex ?)... We would grow socially and intellectually with such unusual friends.

Also, consider how it might be if the extraterrestrials could and would take some of us on a quick trip to their home planet. Through wormholes, that shouldn't take such long times as was earlier discussed. At 'Skinwalker Ranch' there seems to be proof (or at least a huge amount of strong evidence) of a stable wormhole. Again, look on the History Channel for past episodes.

The great series 'Star Trek, Deep Space 9' was about a space station that maintained the location and stability of a wormhole. There is a massive metal object buried on Skinwalker Ranch, which might be just such a station. If so, we are very lucky; we just might, possibly, soon, be allowed to use it-- if we can clean-up our act. Not very likely !

So, who are The Extraterrestrials ? **They are people and they are our neighbors !**

Who are we ?

We are an extremely diverse species called 'Humans'. We are the most intelligent animals on Earth, although, at times, this statement could be argued. Unfortunately, we are sometimes cursed with personal greed, with disrespect of other's rights (often, even the right to be alive ! And if you are an unborn baby, oh my...). We lie, we steal, we cheat, we murder! Why ? Good question, yes ? Could it be poor parenting ? Sometimes it is due to poverty and poor education, probably.   Some of that is due to corrupt governments. Sometimes education is propaganda of hatred rather than acceptance.

We **strongly need** to grow socially as well as intellectually. Obviously, such growth is becoming necessary. There is *no logic* to the conflicts that are constantly cranking up around the globe. There is *no good reason* for starting such conflicts or wars. Every nation should be free of corruption. Every citizen of any nation should have all of the God-given freedoms that the United States of America Constitution guarantees (*somewhat simplified*):

1 Freedom of religion, speech, the press, and assembly;
2 Freedom to keep and bear arms without infringement;
3 Freedom of one's home from government intrusion;
4 Freedom against unreasonable search and seizure;
5 Freedom from self-incrimination in court procedures;
6 Right to a speedy, public trial and counsel, by an impartial jury;
7 Freedom from re-trial after found not guilty;
8 Freedom from excessive bail and cruel and unusual punishment;
9 Freedom from courts creating new laws;
10 Powers not delegated by the Constitution are to the States or to the people.

P.S. Number 2 was not built in for hunting nor for target shooting; it was included for the express purpose for the people to control the governments if they moved toward dictatorships ! The governments are supposed to fear the people who employ them.

These Freedoms/Rights are known as the 'Bill of Rights', signed and added to the Constitution as the first ten amendments in 1791. They are **NOT** given to the people by the government, but are gifts from God, guaranteed by the Republic. The bill of rights is constantly being attacked, squeezed, bent, strangled, misinterpreted, and clobbered by politicians who want total power over the people rather than a Republic. It is always being defended by patriots who understand its purpose, reason, and meaning. Government works for the people, not the other way around. We are the employers !

<u>Who wants to be the first to greet them ?</u>

"Me !" you say, enthusiastically. Yes; me too ! And I, the author of this book, would be very happy to greet The Extraterrestrials. Not that I am qualified enough, but actually, I did meet them in 1977. Although this book is not to be about me, I can and shall relate an interesting and truthful anecdote.

My wife and I, with two young boys were on vacation from California to Canada. We borrowed her father's pickup/camper. On the way back, we stopped at Olympia, Washington for gas and lunch. We enjoyed our sandwiches on the lawn in front of the Capitol building. At exactly noon, we headed south on I-5, intending to spend the night in Eugene, Oregon, about 225 miles. We expected to arrive there at around 5:00 PM.

My wife was driving, to give me a break. I laid my head in her lap and we talked as she drove at 70 mph down the I-5 freeway. It became dark, pitch black in fact. We did not notice the darkness until the headlights went out. She hit the brakes hard-- the headlights came back on. We continued at 70 mph. It was extremely dark. Normal, of course. Nothing to worry about.

Our minds became gradually capable of remembering that the engine died, and that we had stopped at a (lucky) convenient pull-off surrounded by thick forest. Then later, the engine could start. It was completely dark. We got back on the freeway, but the headlights kept going off for a little while. Then we continued down I-5 normally. We didn't think about why it was dark.

A car came up beside us in the left lane and matched our speed. My wife said that there were two young men in the car and they wanted her to stop. We guessed that they thought she was alone and traveling with a bed ! I remained lying on her lap. She saw an exit to a small town named Kelso, Washington. Without warning or slowing, she took the exit at full speed, and lost the two hopeful admirers.

At the end of the exit was a gas station. For safety from the two guys, we pulled in and stopped for fuel. The station manager was a nice fellow. We chatted a bit. I filled the tank-- it only took 4.8 gallons ! I looked at the station's wall clock-- it was 3:00 AM ! I asked the manager if that was the correct time-- it was ! Our watches were also at 3:00 AM ! I discovered we had only traveled 66 miles !! We lost about 14 hours !!

During the rest of our journey back to California, we and the boys did not think anything was unusual. It seemed that we knew that we lost time, but it was not comfortable to think or talk about it. As days past, little by little we were able to think more clearly about the lost time and to talk about a few of the details.

Also, we went shopping at a Safeway grocery store and at the check-out stand, there was a paperback book of science fiction. On the cover was an image of a UFO alien:

**The Captain
of the Flying
Saucer (almost)**

It was the first time either of us had seen such an image-- we both yelled, **"Oh my God, That's him !"**. It looked much like the above but not frowning, not mean looking; our Captain was smiling with his small mouth.

We had a good friend who was a psychiatrist. We told her of our time loss. She was quite concerned, and offered to hypnotize my wife and discover what happened. She recorded wife's memories and wrote them.

We were all abducted by extraterrestrials onto a flying saucer. We were medically examined by a team and robotic machinery. Needles were inserted. After the tests, the Captain of the vessel showed us, thru round windows, what the Earth looked like from a high orbit in space. He showed us around a bit, but not very much of the inside. Then he apologized saying that they would return us unharmed, except that our memories of the incident would be erased.

We did notice that we both had equilateral triangle needle marks on our abdominal areas. Whenever we tried to discuss the incident with each other, we found that it was difficult to think about-- our minds kept turning away into other more mundane subjects. Over the years, tiny bits and pieces of memory returned.

So, who wants to be the first to greet them ? **I do ! How about you ?**

End of Chapter 4

When will we know they are here ?   Page 38
When will we become eligible ?   Page 39
When will we begin to change ?   Page 40
When will we be 'outer space' travelers ?   Page 42

When will we know they are here ?

Most certainly we already know the extraterrestrials are already here. Furthermore, there is plenty of good evidence that they have been here often for thousands of years. Consider one of the oldest mentions of such visits is in the 'Bible'.

Whitney Hopler on Crosswalk.com writes: "In Genesis 1:26-28, before human life is created, God speaks in terms of 'us' planning to create human beings: "Then God said, 'Let **us** make man in **our** image, according to **our** likeness; and let them rule over the fish of the sea and over the birds of the sky and over the cattle and over all the earth, and over every creeping thing that creeps on the earth. And God created man **in his own image**, in the image of God he created him; male and female he created them."

Some readers have suggested that the 'us' to whom God is speaking are other aliens or possibly other Gods !

Hopler further writes: "Genesis 6:4 reveals: "The Nephilim were on the earth in those days-- and also afterward-- when the sons of God went to the daughters of humans and had children by them. They were the heroes of old, men of renown.""

Here is a small portion of the 'Book of Ezekiel' from the King James version of the 'Bible': "And I looked, and, behold, a whirlwind came out of the north, a great cloud, and a fire infolding itself, and a brightness was about it, and out of the midst thereof as the colour of amber, out of the midst of the fire.

 Also out of the midst thereof came the likeness of four living creatures. And this was their appearance; they had the likeness of a man."

There is much more about this experience. For people of that time (almost 1000 years BC) trying to describe a rocket propelled vehicle landing with four extraterrestrials (possibly wearing space-suits), this description was fairly good.

There is an interesting article here:

https://www.christianity.com/wiki/christian-life/should-christians-entertain-the-idea-of-aliens.htm

More biblical references ? The Prophet Elijah was taken to heaven in a space vehicle; his follower Elisha was with him at the time and witnessed this, and II Kings, Chapter 2, Verse 11, reads as follows: "And it came to pass, as they (Elijah and Elisha) still went on, and talked, that, behold, there appeared a chariot of fire, and horses of fire, and parted them both asunder; and Elijah went up by a whirlwind into heaven."

It seems to be a fair description of a rocket powered vehicle, described by a man of zero knowledge about such machines.

Another example: The Prophet Zechariah gave a very precise description of a UFO, in Chapter 5, Verses 1-2: "Then I turned and lifted up mine eyes and looked, and behold, a flying scroll. And he said unto me, "What seest thou?" I answered, I see a flying scroll, the length thereof is 20 cubits, and the breadth thereof 10 cubits." A cubit was approximately 20 inches long, so the flying 'cigar' was about 33 feet long and 16 feet diameter.

Newer clues and evidences are the petroglyphs scratched into rocks hundreds or thousands years ago by indigenous peoples sometimes depict figures that look very much like weird space travelers and/or space vehicles.

And there are those subjects already covered, of course. It seems that the answer to the question, "When will we know they are here ?", is: **Now ! We already know they are here !**

When will we become eligible ?

In America, which seems to have the best foundations for a government that guarantees God-given human rights and freedom, racism is long gone (if only people would quit mentioning the colors of various people). Equality, which truly means the equal right to "Life, Liberty and the pursuit of Happiness", as written by Thomas Jefferson in the 'Declaration of Independence', lives here in The United States of America. Just be sure to realize that we are not guaranteed happiness, just the pursuit thereof.

Most of the people of the U.S.A. should be acceptable to interstellar visitors. But there is a but ! We are a Republic, not a Socialist Democracy. But the constant push by Democrats toward a socialist, Marxist, communist, totalitarianism could easily prevent our nation's acceptance.

As for the rest of the world, it seems that most of the European nations have somewhat reasonable governments and reasonable populations, also.

Those places where there is unrest, dictators, and war mongers, especially extreme religion-based terrorism, are **obviously not** the kind of neighbors that a logical extraterrestrial society would accept as neighbors. We Earthlings should not accept such neighbors, either. Wouldn't it be great if the extraterrestrials helped us get our global acts together ?

How long will the human race put up with the evil denizens of the previous paragraph ? Another good question is; 'Why do we humans continue to put up with such terrible behavior ?' The answer to these questions is the exact answer to the subject question: 'When will we (Earthlings) become eligible ?' **Unfortunately, sadly, and probably-- a very damned long time !**

When will we begin to change ?

We have begun ! Little by little, we are improving. There is a lot of inertia. Just consider that the Republican, Abe Lincoln, freed the black slaves on the first day of 1863 (as was promised in the U. S. Constitution back in 1787), kept from full implementation by another Party, and was finally forced to apply way back in 1964. It was especially difficult in the Democrat South. So, one hundred seventy seven years later is not so good. Also, social racism is finally disappearing after another sixty years. **Total: 230 years.**

Consider the Hitler era, begun in 1933, wherein Jewish people were to be exterminated until they were completely extinct. Hitler and his communist NAZIs were defeated in 1945. **Twelve years ?**

No ! More ! Currently, the Jewish people are being persecuted once again **(and still !)** by religious extremist terrorists and misled college students by some very weird and evil professors. The Communist Muslim terrorists are still in favor of punishments as extreme as stoning women to death for allowing a bare ankle to accidentally be visible. **Ninety years ! Or how much more !**

In order to understand how serious and difficult this question is, here is a fictional scenario:

****

This hyperwarp-space communication was intercepted recently by a Ham Radio Operator in Washington State. She was experimenting with a multiplexed and filtered mixture of commonly ignored noise signals on a hyper-super-ultra high frequency band. The message was sent from Earth toward outer space. A recording was made and then decoded into English with a super-computer using Artificial Intelligence. The process took seven months:

* * *

"FROM: Cloaked Interstellar Command Unit S-9/Q-4 based at Northeast Utah State of The United States of America, North American Continent, Planet Earth, Star System Sol.

TO: Space Monitor Agency via Worm Vortex E-13

SUBJECT: Report of 60+ Earth years observation in expectation of offering membership to this species of animals known as Humans:

BEGIN KNOWLEDGE DATABASE DUMP:

The Arab terrorists, 'Al Queda', a broad-based militant Islamist organization founded by Osama Bin Laden in the late 1980s was mostly gone by 2021. Some of 'Al Queda's power has been distributed to other terrorist groups, all of which have the same driving force: religious fanaticism. **Forty years ? More !**

ISIS was a very similar terrorists organization, formed in 2004, one of the offshoots of 'Al Queda'. On December 19, 2018, President Donald Trump declared that ISIS was defeated. **Fourteen years !**

According to the United States Central Intelligence Agency (CIA), there are sixty significant terrorist groups. The current terrorist groups in the news that are attacking the nation of Israel are: Hamas, Hezbollah, Islamic Revolutionary Guards, and Palestinian Islamic Jihad. The current war, wherein Israel is defending itself from attacks from all directions, **has been one year at this writing.**

Also, there is a war going on in which the nation of Russia wants to re-assert its claim of ownership of the nation of Ukraine. Communist Muslim Democrat President Obama of the nation of United States of America allowed communist President Putin of Russia to infiltrate and annex part of Ukraine (Crimea) in 2014. **Ten bloody years of war has ensued and is continuing.**

The Earthlings seem to be handling these negative happenings a bit faster lately, but still quite poorly. Perhaps that means they are becoming a bit more intelligent and more peaceful. When will they begin to change (significantly enough to matter) ? **Not yet-- but soon, we continue to hope !**

Some of these humans have the proclivity to cruelly, slowly pound to death with rocks-- another human, a female who is standing in a pit, defenseless and hopeless. And for a silly, useless, nonsense reason-- a stupid, illogical belief system which the victim had no choice about !

RECENT EXAMPLES: Into the peaceful nation of Israel, 6,000 Islamic terrorists known as Hamas, from a territory known as Gaza, 3,800 Nukhba forces, and 2,200 civilians surprise-attacked military and civilians alike. The attackers raped and/or killed many, totaling 1,139 dead humans, and injured thousands more with many hundreds of rockets.

Our mobile cloaked outpost insisted upon interjecting power to stop the meaningless terror, but we, at the Command Unit S-9/Q-4, would not sanction an operation due to the Prime Directive of non-interference in the affairs of primitive beings.

A different example is the ongoing incursion of the South Border of the United States of America. Human trafficking, including child abduction for sexual purposes, and slavery is rampant. Horrid addictive drugs are also smuggled into the USA. That border was closed under their past President Donald Trump, a Republican in political terms, but has been completely open with governmental help by the current Democrat President Joe Biden and his 'border czar' Vice-President Kamala Harris.

SUMMARY AND CONCLUSION: It is abundantly clear that this 'civilization' is **not yet ready for open contact**, even though their technology is quite good and improving quickly. We still have no estimate as to when this 'civilization' will realize their social faults and remove them.

END OF TRANSMISSION

* * *

End of <u>fictional</u> scenario.

****

<u>When will we be 'outer space' travelers ?</u>

That question refers to space beyond our solar system; in other words, much longer distances such as interstellar travel, not interplanetary. Going to another star and its surrounding planets, as examined previously, is extremely difficult.

Given the discussion in the previous question, maybe we must give up hope of joining our interstellar neighbors entirely.

Any sane extraterrestrial would watch from a distance in a 'cloaked' vehicle, then report back home: **"Stay the hell away from these primitive beasts for another thousand of their Solar years !"**

So, when will we be 'outer space' travelers ? Ever ? **Not bloody likely !**

End of Chapter 5

Neanderthals ?
Image courtesy Yulia Serova

Page 42

Chapter six   Where ?

Where might we find the extraterrestrials ?   Page 43
Where might we find evidence of them ?   Page 44
Where should we look deeper for evidence ?   Page 45
Where would they come from ?   Page 45
Where would immigrants from planet X live ?   Page 46
Where will we be on the interstellar power chain ?   Page 46
Where would funds come from when we join ?   Page 46

Where might we find the extraterrestrials ?

Underwater ! There are instances of military pilots who have reported seeing UAPs diving at hyperspeed into the ocean. If that is true, underwater is one place to look. Somewhere southwest of Los Angeles is a hot spot. Off the Atlantic coast is another possible search area. Here is a great government video of UAPs, some seem to be entering the ocean:

https://www.cnn.com/videos/business/2021/05/19/ufo-navy-video-jeremy-corbell-orig-jm.cnn

Earth is about 71% covered by oceans. Some parts of these waters are deep and other parts are extremely deep. Much of this underwater has not been well explored. Who knows what an underwater base would look like ? Of course camouflage would be easier underwater, especially for advanced people such as extraterrestrials. The privacy they would enjoy would be superb.

Underground ! Again, easy for them to do and to camouflage. There are several places that are currently being investigated. Obviously, Skinwalker Ranch (previously mentioned) is the best place to look. There is an abundance of wonderful evidence there that a huge metallic object is buried, just next to the probable 'portal to a wormhole' that the researchers have discovered.

Searchers are also finding evidence of similar places in Colorado, Arizona, and more. Rumors of underground cities are still circulating. Former Governor Jesse Ventura investigated some of them several years ago (2012) on TV. He did find some that were, indeed, extremely odd-- and huge. For more information about his interesting series, look here:

https://duckduckgo.com/?
t=ffab&q=jesse+ventura+conspiracy+theory&ia=web

The most popular conspiracy theory has been the Denver International Airport, but it seems that nothing definitively alien is underneath DIA. If there is something hidden deep beneath there, it is very well hidden, since Jesse did not find it.

In the open with rubber masks ? No, probably not. Shape-shifting ? No. probably not. Possibly ? Yes, of course. A technologically advanced civilization, such as extraterrestrials, could do many such visual effects quite easily, without breaking any laws of physics (**there is NO magic !**).

If a person that you know of was an extraterrestrial, would it be able to appear, move, walk, eat, breathe, speak, dress appropriately, play golf, sit in a sauna and sweat, make love, get drunk, etc., but never once slip up from appearing human with proper weight, smells, stride... ? Probably not. Would it be that helpful for the extraterrestrials ?

Hovering in the sky, cloaked ? Once again, easy for them to do and to cloak, just as in the 'Star Trek' TV series. Our own government, as well as private companies, have been working on these techniques for years. Here is an interesting website to see some actual, real cloaking devices work, that are made by Earthlings:

https://www.zmescience.com/feature-post/technology-articles/inventions-1/real-life-invisibility-cloaks/

At Skinwalker Ranch, it is obvious that an object is hovering overhead that is entirely cloaked. Maybe it is ours, but it seems to be much too advanced, so it is most probably the extraterrestrials. So, in summary, plenty of places to look for the extraterrestrials, but looking for them and actually seeing them are not the same thing.

Where might we find evidence of them ?

Evidence of past and present visits to Earth have been found in writing from biblical times and petroglyphs (as discussed earlier). We can certainly find huge amounts of evidence held by the U. S. government, as well as that held by other governments around the world. Americans are not the only UFO/UAP researchers.

SETI, the Search for ExtraTerrestrial Intelligence has been listening for radio signals from outer space for a lot of years. Sadly, they have not found any provable, repeated evidence. We should continue this form of search with more backing and vigor.

A great place to find evidence would be on our moon, Luna. With no atmosphere, footprints take eons to be erased by dust from impact craters. We might find something there that is not natural.

Mars is another place that might have been visited by extraterrestrials. The winds on Mars can be brutal, though. NASA writes: "Mars is infamous for intense dust storms, which sometimes kick up enough dust to be seen by telescopes on Earth. The winds in the strongest Martian storms top out at about 60 miles per hour." The reason is that the Martian atmosphere is only one percent as thick as Earth's. The problem with those winds is not the speed; it is the fine dust, which sometimes swirls around the entire globe.

Any evidence that Mars was visited by extraterrestrials would be erased unless we could find protected places (caves ?) or underground places.

Some of the moons of Jupiter or Saturn might be places to search, but these are so far away that we can't reach them with people yet. Probably soon we can get a robot there. We will at some point. And Mars has two small moons that are within our reach. Like our moon Luna, they have no atmosphere. Evidence will stay ! **Let's go look !**

Where should we look deeper for evidence ?

As mentioned above, but we really should rent a D-11 bulldozer and dig the top off of that mesa on Skinwalker Ranch ! There are several other places that need the same brute force methods to uncover hidden stuff. One is in Nevada near Site 300. Another is in Colorado. There are more...

We should trawl the oceans around the known hot spots with instrument packages that can detect electromagnetic field strengths, magnetic field changes, temperature anomalies, radiation, ultra-sound, sonar, various light frequencies, and whatever other stuff there is. Chemical analysis of the water for anomalies should be a priority.

We need to figure out how to defeat (or at least detect) the act of cloaking. Then we should search the skies with constant scanners, looking for patterns or anything unusual, then record them.

What if we find them ? What should we do ? We must put our best foot forward, as some people say-- then **try to convince them that we are friendly and worthy !**

Where would they come from ?

Given distances to nearby stars, we might limit that answer to those closest to us. But not necessarily, of course. Further limit it to those star systems that have 'Goldilocks' planets in 'Goldilocks' orbits. Those planets that are similar to Earth, which is currently THE ONLY **KNOWN** PLANET THAT **SUPPORTS LIFE !** Probably, those limits are too stringent, given that our knowledge of interstellar life is nil.

Given wormhole reality, only God knows how many wormholes there are, where they start and end, and how lengthy or short they might be. Logically then, the number of possible visitors and the places they live are huge. A lower limit number would be from 3 to 7 of the nearest stars. A reasonable upper limit might be the nearest hundred or stars. Just logical guesses, of course, but...

They could come from anywhere !
Where would immigrants from planet X live ?

If the extraterrestrials were to immigrate here, would they simply move into the house next door ? Would they be so different that they might require a chunk of land in the Arizona desert or the Southern California desert ? Or might they require a piece of Washington State forest ? What aboot Canadar ? Eh ? Antarctica ? Tropical South America or Africa ? How about a South Pacific atoll for quarantine purposes ?

That cluster of questions illustrate the point about their air and climate needs and, possibly, their physical resilience and their normal germs (if any). Maybe, if they need very different air mixtures and/or pressures, a sealed outpost could be the only solution-- and that would mean just for a small number of extraterrestrials.

What about living space ? If you think that America is full (or too full), just look out the window the next time you fly across the country. We have gobs of emptiness ! America is not alone; most nations have huge amounts of empty land. Of course, not all of it is habitable.

If they are water dwellers, *oh my* ? Just look at the globe; Earth is three fourths water ! Dolphins and whales would have some new neighbors. Suppose they can't stand UV solar radiation (AKA sunshine). Perhaps underground cities could become a reality. There are some problems if a huge number of extraterrestrials moved in-- infrastructure, utilities, jobs, food, all of the usual. **But, living space is not a problem !**

Where will we be on the interstellar power chain ?

If Earthlings are allowed to join the interstellar or intergalactic community, you can rest assured that, as the latest newcomers with the least technology, and given the problems listed in Chapter five, we certainly would be at the very bottom !

Can we Earthlings ever climb higher on the interstellar importance pole ? **Who knows ?**

Where would funds come from when we join ?

For the U. S. A., consider our Constitution; our Federal government was supposed to get its funds from trade tariffs and fees, not from taxes. States were allowed to levy taxes. Congress has screwed that up terribly ! Other countries around the world impose taxes, tariffs, and fees, just as we do.

Currency was (one uponce a lery vong time ago [silliness intended]) based upon rare and expensive metals such as silver and gold. That went away. Precious metals aren't so rare anymore, anyway. Who knows what our currency is based upon now ?

Maybe the extraterrestrials need water. We have a bunch ! Maybe we need medicines and they have a bunch. We could exchange lots of ideas. We could trade goods. Our plants would be different from theirs, and vice versa. Some of those plants are the basis for medicines. Perhaps their plants are the basis for their medicines as well. Cross-comparing medicines could become a hugely important source of funds, assuming they use money.

As for coin, currency, cash, money, moolah... perhaps it will become obsolete. Consider once again the Star Trek series; those characters of all sorts of shapes, from all sorts of planets seemed to have no need for money.

For nostalgic purposes, here is a familiar image:

Starship USS Enterprise NCC-1701

Everything was free because the replicators would respond to verbal commands and instantly supply the "Tea, Earl Grey, Hot." as well as almost anything else. If everything in their lives was free, the implication is that **energy was free and plentiful !** Probably, Star Trek's energy came from Laser Fusion.

Remember, energy and mass are equal. Can we figure out a method to carry monstrous amounts of energy into space instead of monstrous amounts of mass that is used to produce energy ? Energy doesn't weigh very much.

Energy is the most important part of everything that is built or made or produced. Second is labor. Perhaps robots in the 'time' of Star Trek did everything for everyone. Still, there was the crew that worked on the Starship USS Enterprise NCC-1701. Did they get paid or did they work because they liked doing what they do ? The implication is a form of communism. **Ouch !**

Communism has sounded so wonderful to many cultures and so many times, but it has never succeeded in the long term unless it is imposed by brutal dictatorship. "Why not ?", you ask. Simple-- every time it has been peacefully tried, (communal work, communal sharing), there are those who want to share the goodies but do not wish to share the work. The communal garden becomes overgrown with weeds and undergrown with veggies. Also, some of those people want a bigger share than others. Usually, such bums want a huge bunch more ! **Who decides how much ? A dictator with brutal force !**

There is one exception; family. Communal living works for families because there is the bond of love. Love is the power that can overcome the laziness that follows from an unending supply of free food, etc. Otherwise, tyranny provides the incentive to work, with the bare essentials free for almost all, but luxurious plenty for those in power.

Maybe... *just maybe*... if unlimited and free energy is readily available to all, and robots did all of the dirty work and heavy lifting, everyone could become their ultimate self: doctor, lawyer, Indian Chief, Star Trek Captain...

The lazy could be lazy if they wished. Maybe no one would care. However, social pressure might provide incentives... probably not.

Communism will not and cannot work at the present time ! Probably, hopefully, it never will.

Taking wealth or goods from those who have it and giving it to others (Marxism) is simply a form of thievery. However, if energy is free in unlimited quantities, and robots do a lot of work, then wealth would have little meaning. This all leads to an abhorrent thought: some form of money-less communal global government, without dictatorship, might be viable in the far future. *Shudder...* **Who knows ?**

End of Chapter 6

Chapter seven   Which ?

Which types of extraterrestrials were here long ago ?   Page 49
Which types of ETs are here now ?   Page 50
Which types of ETs would likely immigrate here ?   Page 50
Which types of ETs would not immigrate here ?   Page 50
Which types of ETs would be fun neighbors ?   Page 51
Which types of ETs would you not like ?   Page 52
Which kind 0f planets would we likely emigrate to ?   Page 52
Which kind of Earthlings would likely emigrate or vacation ? Page 53

Which types of extraterrestrials were here long ago ?

There are at least three major sources of very ancient visits from extraterrestrials. One is the writings of religious texts such as the Dead Sea Scrolls. Consider Ezekiel's description in the Bible of a fiery chariot carrying four strange beings. It seems that he thought of that sight as one being with four faces. They moved as one in a vehicle with a wheel within a wheel. That could be a sealed rover-like vehicle with the occupants each looking a diffferent way. Sealed because of a different atmosphere for the space travelers. He said the vehicle didn't turn; it went sideways and back and forth.

A good guess might be that this type of extraterrestrials were only slightly past our own technology, such as our moon landing in 1969. *P.S. This author did the first physics, math, and binary machine language programming for the moon landers tine computer in 1963 ! And the answer is: "Yes, we actually did land the first two humans on the moon.*

Ancient petroglyphs that depict beings which look like extraterrestrials are found all over the world. Their ages are different and difficult, but they are very old. Some of them seem to have glass helmets. Spacesuits, maybe ? Perhaps these extraterrestrials were a little more advanced than Ezekiel's. They were not staying in a 'pope-mobile' with enclosed glass; they walked free with just a helmet.

At Skinwalker Ranch in Utah, as mentioned before, there is what seems to be a huge metallic vehicle buried under a large mesa. So many strange happenings and so many types of energy radiating from underground implies strongly that the machinery (at least) is still operating. The mesa looks to have been disturbed by non-natural means, yet is long well-weathered. Native Ute and Navajo tribes have old oral histories of odd beings and weird happenings and portals. The portal that seems to be there is still active, as is the cloaking, which implies more recent extraterrestrials still visiting. This could date the 'vehicle' to an era a few hundreds of years ago, and imply much more complex technology. So, the answer is ably several types, not too weirdly shaped.

Which types of extraterrestrials are here now ?

Certainly, those extraterrestrials ,which are now here, are the most technically advance of the visitors. Obviously, they are peaceful, curious, and friendly. If they were not, H.G. Wells's 1898 novel, *'The War of the Worlds'* would have already happened. It has not !

Plenty of people who have been abducted say that there are three major types, maybe more. They are described as the image on page 37 most commonly, and smaller, darker, more robotic-like beings, which do precise surgery and other tasks. Also a larger, more human looking type has been reported, which seems to be interested in cross-breeding experiments. Reports are that they are very gentle, polite, and quite sexy. There are rumors...

Which types of extraterrestrials would likely immigrate here ?

Immigrants to another country are quite often younger couples, off to explore a different life style. Sometimes they are looking for better opportunities, or at least, different ones. Perhaps business opportunities would provide their motive.Maybe they want adventure. Possibly, they may wish for their own place in the history of two worlds.

Naturally, they would wish to breathe freely of our atmosphere rather than enclose themselves and produce their own type of atmosphere. That fact might limit or even preclude any actual permanent immigration. In other words, your new next door neighbors will breathe our air.

Some extraterrestrials which have ambassadorships or long-term government positions would likely immigrate to Earth, at least for awhile.

But, other factors might also preclude permanent immigration, such as types of germs and diseases, although their advanced technology would probably keep them safe from our diseases and hopefully vice-versa.

Immigrants which may move next door would be outgoing, adventurous, brave, curious, and very friendly. Which of us would not like that ? Imagine the neighborhood parties, street dances, barbecues, and card games. Exchanging stories, jokes, recipes, food, languages, histories, knowledge... more fun than a barrel of monkeys. Are you curious ? Are you brave ?

Which types of extraterrestrials would not immigrate here ?

Hmmm... Given the state of our world (see pages 30 and 40), probably most extraterrestrials would not wish to immigrate to Earth.

Certainly, any extraterrestrial who is happily content with their current position and life would rather stay home and watch TV shows of Earth and

Earthlings. Those extraterrestrials which are elderly would probably prefer to stay home in comfort and ease, just as we would. Any of them which might have more difficulty than others enduring our weather, our atmosphere, or possibly even our sun's color spectrum would not come to Earth on a permanent basis.

The unknown (to us, currently) hazards and discomfort of interstellar space travel may preclude many immigrants. Furthermore, any of us which have moved to another country, state, or city know that we carry a lot of baggage. Furniture, for example. How much personal goods, which we have accumulated, **must** be moved with us ? Can it all be moved from planet X to Earth in an interstellar spaceship ?

Any extraterrestrials which do not want to leave such goods behind will not move to Earth.

Which types of extraterrestrials would be fun neighbors ?

That question requires some vivid imagination ! Since very little (if any) real facts are known about their personalities, we must imagine a personality that we would appreciate. For an immigrant couple, or family, just look around. The same set of personalities that we now enjoy as neighbors (if we are lucky), would be perfect.

Very outgoing would be great. Traveling this far to a very strange land would certainly imply an outgoing extraterrestrial.

A good sense of humor would be much appreciated. Laughter always draws people closer. Plus, it makes for good friends. If they laugh ! Surely they must laugh, don't you think ?

Those extraterrestrials which are courageous would be a positive trait for a new neighbor, and, of course, traveling here from planet X would certainly require some large amount of bravery. Such neighbors would not be shy.

Those neighbors which are talkative are usually more fun than the quiet ones. Naturally, language would be a temporary obstruction for both sides, but the process of learning in both directions has always proved to be fun, comical, sometimes embarrassing, and always conducive for closer friendships.

Looks and appearances sometimes alter a persons attitude toward neighbors. However, in the case which we are considering (extraterrestrials), we cannot allow appearances to interfere with our acceptance or our feelings. That also works both ways, of course. If they appear to us as a bit reptilian (a popular themed idea, which is probably untrue), some of us Earthlings might have difficulty getting past the first few meetings. Hopefully, we can manage such a conflicting feeling, eventually.

Smells could be a problem which we (and they) may (or not) find offensive (or really nice !). Rumors are that some have an odor of cardboard, but that might not be true.

So, the answer is variable until we meet.

Which types of extraterrestrials would you not like ?

We had better keep this part a secret from the extraterrestrials. We do not want to piss off new neighbors ! This should be short and quick to answer. Clearly, it is all of the opposites in the above question. But nobody likes the person which hogs the conversation. Nobody likes people who know-it-all. Nobody likes to be close to the infamous B O. No one likes farters either.

On second thought, maybe we should leak the above paragraph to them. On the other hand, the extraterrestrials probably have a similar list to print up and hand out to us. It will probably be much larger !

Which kind 0f planets would we likely emigrate to ?

A subtropical paradise ? Not always. Some of the subtropical paradises are extremely humid. Some of them have consistently strong winds. Many of them have denizens that are far less than pleasant, from snakes and crocodiles to jaguars and huge hornets. And that's right here on Earth !

Many people like to ski. Snow can be nice. Others hate snow. Rain is good. Floods are not. Four seasons please a lot of people. Of course, you knew all of that. But...

Which unusual features make a place exotic ? How about two moons ? Which of our own planets have multiple moons ? Yes, several. Since some of our own solar planets have more than one moon, we can assume that most star systems have planets which have multiple moons as well. How romantic would that be ? Also, multiple moons usually have different orbits and speeds, so moon-rises would be random and probably never repeat.

Some star systems, like Alpha and Proxima Centauri are doubles, which truly complicates daylight and dark for any planets which orbit within that system. Weather would be more complicated for any being than a sudoku problem that starts with only three numbers !

The answer to this big question is (for most of us) a planet which we call a 'Goldilocks Planet' would be the best choice. However, multiple moons are still an exotic difference, which would only cause weird tides.

Which kind of Earthlings would likely emigrate or vacation ?

My kind ! Your kind ? Considering what data we have, it seems that the extraterrestrials have somehow managed to control gravity. If that is true, then the trip would not be uncomfortable. Also, it seems that they have managed to use wormholes. If that is true, then the time of travel will be tolerable.

It is quite clear that they are not enemies, so the only remaining problem for the extraterrestrials to give us a free ride to their planet would be the atmosphere, diseases, and weather. Certainly those are things we (or they) can control, at least for a short time such as a vacation. Emigrants from the good, green, watery Earth could possibly have fewer choices in planets.

Neverthemore, the idea is very exciting. Which of you readers (if you aren't too old) are ready to try ? You don't have to be brave hero explorers like the Vikings or Christopher Columbus, since science has made it easy. Our younger generations should certainly be seriously preparing and looking forward to **full contact. It will happen, later or sooner !**

A scene from the great TV series 'Vikings'

End of chapter 7

Chapter eight   Why ?

Why do so **few** people believe there are extraterrestrials ?   Page 55
Why do so **many** people believe there are extraterrestrials ? Page 55
Why aren't animals with larger brains smarter than us ?   Page 56
Why are extraterrestrials so much more intelligent than us ? Page 57
Why can't we find a warp-speed method of travel ?   Page 57
Why are extraterrestrials here ?   Page 58
Why are we here ?   Page 58
Why did this author write this book ?   Page 60

Why do so few people believe there are extraterrestrials ?

That's a very difficult question to answer. So, let us make a few guesses.

A – Fear. Some folks fear that if they accept the reality of extraterrestrials existing, then they will have to worry that such aliens, smarter than humans, will mke us feel shame, or take control over us, or maybe even murder us whilst we sleep. For those such people, disbelief feels much safer.

B – Misinformed. Given the large number of fake reports, and the fact that governments lie, deny, and hide the truth, belief is difficult.

C – Disinterest. Many folks are not interested enough to research the subject. If they were to do a small amount of research and filter out the ridiculous fakes and learn the small amount of easy physics to see that it is possible, then they might get interested.

D – Religious fanaticism. There are some religious folks who don't wish to share their god with off-world beings. Or some might say that, if it isn't in the Bible or other writings, it can't be true.

E – Ridicule. Some will ridicule believers and call us "aluminum foil hat nuts." Nobody wants to be ridiculed, unless they are actually ridiculous. Then they should be ridiculed.

F – Ignorance. Ignorance is not a sin. Ignorance is not stupidity and it is fixable. You can't fix stupid.

To sum it up, some of each keeps some folks from believing.

Why do so many people believe there are extraterrestrials ?

Quite a large number of people have seen UFOs or UAPs. Many people have had 'lost time' experiences. More than a few have been abducted and retain some memory of the abduction, including this author.

This author had a previous experience with a UFO in the fall of 1957 in Southern New Mexico. Although it did not involve seeing extraterrestrials, as in the Washington State abduction in 1977, it did involve a flying saucer and a number of regular people on a lonely highway.

My first wife and I were driving from Alamogordo past the White Sands Missile Range to Las Cruces. There was very little traffic that day. The car's engine simply stopped as a flying saucer settled down just ahead of us, off the highway about 40 feet. It was about 20 feet across and maybe twelve feet high in the middle. The saucer seemed to be shiny metal with a few round windows.

There was no place to pull off the highway. We coasted to a stop on the road and looked at the saucer in awe, but rather calmly, which was **very** odd.

Soon, other cars coasted to a stop behind us until there were perhaps ten or twelve cars. We got out and other occupants of the other cars got out. We all stood around on that highway and calmly discussed the fact that there was a UFO beside us. No one seemed surprised. No one panicked. It seemed to all of us only mildly unusual.

After ten minutes or so, the saucer slowly, silently lifted off and rose into the sky until it disappeared. All of us on that highway just got back into our cars and drove onward. During the rest of the trip to Las Cruces we found it more and more difficult to discuss the details of the unusual happening. We simply stopped talking about it for several weeks. Eventually, we were able to discuss the basics of the incidence, but details are still impossible to recall.

Huge numbers of people have seen or read about the massive amount of evidence of the extraterrestrials. Many scientists know the physics previously discussed that strongly suggests the possibilities and the probabilities of their presence. It is quite likely that many of us humans simply believe the extraterrestrials can exist because the 'Star Trek' series made it look very likely and possible. Furthermore, a bunch of us Earthlings believe they are here (or at least exist) simply because they wish for that fact to be true. Finally, a few people have actually seen them !

Why aren't animals with larger brains smarter than us ?

In other words, since whales and dolphins (and maybe other big animals) have bigger brains, why are we smarter than they are ? And the following implicated question: "Does that mean that extraterrestrials have bigger brains and therefore larger brain-boxes ?"

Who said that dolphins are dumber than us ? Based upon what ? That brings up some more thought about the evolution of intelligence. And what do we mean by 'smarter' ?

Dolphins and other such sea mammals live in their environment comfortably, else they would have moved to better sea-climates or evolved differently. They eat all they want without much manual labor. They have sexual encounters frequently without shame. They play and sing together and seem happy. They treat their children well. They don't wage war on each other. They have no druggies, beggars, taggers (that ugly graffiti), or homeless.

Their is no poverty, no money thievery. What more could they want in their Eden ?

So, there was no reason or need to cause dolphins to develop any more brain-power for technology, even **if** they had tool-handling ability. Perhaps they use their larger brains for enhanced good feelings about their lives.

We humans, on the other hand, were relatively weak and fragile. Our environment had wide temperature ranges, making us very uncomfortable. We had no polar bear fur to keep warm, nor fire for a long time. We were stalked by larger animals as food. We had to learn how to make weapons and hunt.

Our evolution was forced by danger, discomfort, and need. Consequently, we became competitive problem solvers. Hence, our 'higher' intelligence. Sadly, we also became selfish due to food difficulties and weather difficulties. That might explain our problems as a global community. Evil does exist !

So, some extraterrestrials might or might not have fatter heads and bigger brains than us. Maybe they evolved farther with dolphin-like brains-- the best of both brain types !

Why are extraterrestrials so much more intelligent than us ?

Maybe they aren't ! But they probably **are** smarter, due to a longer time of evolution, which equates to better understanding of the laws of physics and better technology. We are still evolving. Notice how much larger they are making Earthlings nowadays ! Only a lifetime ago, people were shorter and far less weighty than today.

As for smarts, our evolution of intelligence seems to have slowed, stopped, and is going into a reverse mode. Is that actually evolution reversing ? Or is it just piss-poor teaching caused by political forces ? **Well... Yes !**

Consider the reading comprehension of United States high school graduates now, compared with a lifetime ago. It is abysmal ! Math and Science are way down from what it has been and from much of the rest of the world. Furthermore, world wide, our graduates are way far down on the 'totem pole' in every subject. We were once very near the top. College students are being taught some of the most stupid ideas ever thought of, and of course, they believe it. They paid for that 'knowledge',-- or did **you** pay for it ? **Yes !**

A gentle reminder: the United States Constitution makes no mention of a school system. The tenth amendment should give control of schools to the States or to the people. The Feds have no good reason to be involved in education. We should remove the Department of Education entirely !

Why can't we find a warp-speed method of travel ?

Suppose there is a phenomenon called a tiny black hole ? The famous physicist Stephen Hawking theorized that they might exist. **And they do !** If a spaceship aimed itself at a small black hole that was moving at a huge speed, the spaceship might be able to maintain a safe orbit distance and be pulled along by the black hole. Then very high speed would be achievable.

If, in your high speed hitch-hiking travel along with one black hole, you find another one going in another direction, you could switch hitches and continue exploring even farther and faster ! It would be similar to tacking with a sailboat. Maybe that is 'warp speed'.

Why are extraterrestrials here ?

See "What do they want from us ?" on Page 9. Also, consider this question: What if the extraterrestrials are simply **human**itarians ? (oops, can't be; *altruistic* was meant). Maybe they don't need anything. Maybe they just enjoy being nice or helpful. Maybe their religion requires them to be good to their fellow **human**. (oops, can't be; *being* was meant). So, the question: "Why are they here ?" is best answered by: **"Because they can !"**

Why are we here ?

In the following paragraphs, there may be more unknown cases than shown.

We are here ! That is a fact. Where are we ? On a planet named Earth, orbiting a star named Sol, which is a part of a galaxy named Milky Way, hurtling through our Universe. Those are facts. So, to tackle the question of why we are here, we need to consider another question: How did we get here ? And then, to tackle *that* question, there is yet another question to be considered: First, how did our Universe begin to exist ? Well, Sherlock, there are only two possible ways, choose the right one:

A – It randomly happened naturally, obeying the laws of physics.

B – It was caused by our God, obeying the laws of physics.

For the "A" case: It took an unimaginable amount of energy to become a universe during the Big Bang. Energy cannot be created or destroyed, according to physics. Where did it come from ? It was not here about thirteen and a half billion years ago. Who originally created the energy (which cannot be created) ? Who brought the energy to the Big Bang? And who designed the laws of physics so precisely ? A random accident ?

For the "B" case: Our God told us that he created everything. That could include the laws of physics, although they probably exist in God's universe too. An all-powerful God, living in his universe, could direct energy to an empty space, causing the Big Bang, thusly creating our Universe. No magic; no accident.

The "B" case is more logical, since the "A" case requires a supreme being anyhow.

Next, let us wonder why our God created our Universe. There are only a few possibilities:

C – God was forced to by his boss.

D – God was playing with his toys for fun.

E – God wanted a place for us to live before he created us.

F – God accidentally discharged energy into an empty space.

G – Some other being did it and claimed that it was our God's doing.

This author rejects "C", "D", "F", and "G" for fairly obvious reasons, leaving only case "E".

Next, let us wonder just how we arrived. Again, there are only a few possibilities:

H – Magical instantaneous creation (rabbits and top hats method).

J – Handmade of mud, with magical breath to make us come alive.

K – Through natural processes over time using physics laws (evolution).

L – Some powerful being held us in hibernation, and dropped us down.

M – We were abducted from another planet and placed here.

This author rejects "H", "J", "L", and "M" because they don't have much basis in reality, or else they require 'magic', which does not seem to exist. They also require the ignoring of ancient writings.

Remember – there is no magic; there is only our God's physics. Time is relative and not so important to a supreme being such as our God. Why **have** the laws of physics and then not use them ?

So, why are we here ? **What if:** It is because our God wanted us to be here, and to test us for our goodness and badness, then to see if we could fix our badness. Unfortunately, evil does exist, and somehow, it infiltrates minds.

Our God told us that he created us as written in ancient scrolls. He gave us some commandments that would promote our happiness. He wanted us to be happy. With such a huge Universe, no doubt there are other life forms under the same commandments, working to fix their badness. He also said he wanted us to join him in a place (probably in the space-time continuum) called Heaven.

He even said that if we screwed up, repented sincerely, and asked for forgiveness, he would forgive because he loved us. Then there was Jesus. Many people believe what Jesus said. He promised that if we believe in him and ask for forgiveness sincerely, we would get to Heaven. Not a bad reward, yes ? What is so complicated about these two parallel beliefs ? It seems reasonable to believe that it is the final push of our evolutionary journey.

This belief system is simple, not radical, not difficult, and has a plethora of evidence for correctness, including the Dead Sea Scrolls. And for those atheists who say that our God cannot build all of this in six days, it depends upon his view-space-- if he was riding a wave at the outer edge of universe expansion at near the speed of light ('c'), then his clock moved extremely slowly whilst Earth's clock ate up 13.7 billion years ! What does any being have to lose ? **It can't hurt !**

Why did this author write this book ?

You, the reader, have probably noticed that this is not your ordinary book. This author has tried to refrain from injecting his own personal happenings into your reading material. But, finding that impossible, in addition to the two incidents which I have already written about, there is more that I now feel obligated to relate...

In 1978 on a Wednesday evening I was digging some more on a swimming pool that I was building in the backyard. A tough bit of dirt required an extra push, so I put the shovel handle on my (then) hard, flat tummy and pushed. I heard and felt a slight "pop".

The next day at work, I didn't feel so great. Coffee did not seem to stop on the way down. It continued downward and felt hot inside. Friday at work, I began to hurt a lot. In the afternoon, I went home. My wife drove me to the local hospital emergency department. That small hospital had few technical facilities, and x-rays showed nothing, but they kept me in bed for observation.

Saturday, there were no doctors at the hospital. I waited. Sunday morning, an elderly Chinese man was the only doctor there. He said he was a recent immigrant doing residency to get his credentials. He looked at my vitals data. He said that we could not wait any longer, something was wrong in my abdomen and he was going to do exploratory surgery.

After surgery, when I woke up, the old doctor was there beside my bed. He spoke in broken English: "I have good news and bad news to you."

I asked, "What's the good news ?"

He replied, "I find you probrem and fixed it. You appendix bursted."

I then asked, "Oh ! So what's the bad news ?"

"You must call you famiry; you going to die."

I responded, "If you fixed it, why will I die ?"

"When you gangrene in finger, we remove hand. When you gangrene in hand, we remove arm. You are gangrene in belly. Cannot remove you whole abdomen. So sorry. Cannot save."

I asked, "What about antibiotics ?"

He answered, "You allergy of Penicillin, but we give you anyway. Even so, too much gangrene, too much tissues damage. Call famiry.'"

I called my wife. Later that afternoon, I died. My first impression was that the horrid pain was gone entirely. What a delicious relief! Then I began to realize that I was no longer in my body. Next, I became aware that I was truly dead. There were a few moments of extreme sadness, followed by a realization that there was nothing that I could do about it.

Then I became aware of a very bright light in the upper left corner of my awareness (vision? Not exactly). There was a strong pull from that light. I floated toward it, and my entire life flickered through my mind as if a super-fast audio/video played. I wondered in awe as I floated upward, how that could happen with my full comprehension in such a short time.

My floating upward speed increased rapidly. I was rocketing toward that light! I began to feel the most wonderful feeling that I have ever felt. It was euphoric! I reveled in that marvelous feeling as I got closer to the brilliant white light.

Voices began to appear in my mind. At first, it was like muted crowd sounds, indistinguishable. As I got closer, the sounds became distinct. The voices were welcoming me, telling me to be not afraid, that all was well, that my euphoric feeling was normal. I began to ask questions: "Who are you? Where am I? What should I expect?"

The answers were: "We are residents of this wonderful 'space'. You are almost here now, also. You can expect to be here in happiness with that euphoric feeling forever. There is no reason to hurry. Here, time has no meaning."

"I'm dead! Aren't I?"

"In your terms, yes. You still exist... your soul."

"Are my parents here? Relatives? Dead friends?"

"Yes, many."

"Why are they not here for me to see? And why do I not see you?"

"Your mom and dad are currently busy. There will be an eternity to visit with your parents and friends. You are not quite at the end of your travel."

"Are there extraterrestrials here? What do they look like? And what will I look like there, since my body is not with me and was in poor shape?"

"Yes, of course, silly boy, there are 'beings' here from other planets... *Polite laughter.* You didn't think that, in this great universe, you, from Earth, are the only sentient beings. Visual appearance here is as you wish it, since there is only energy here rather than matter. You will soon see and understand."

"Oh... That's good news... But, there are some very important things back on Earth that I need to do. May I please go back to finish, if I can return here?

Many soundless whispering voices in my mind. Then a single clear voice: "You will be allowed to return to your life to finish your job. Make the remainder of your life a proper one. You will return here if you can."

Before I could think the words, "Thank you," I was floating toward my cold body on an operating table with the old doctor and some nurses standing around. The brilliant white light was not there. I slipped back into my cold, slimy body, did not like the feeling, but soon it felt normal. I woke up to a surprised bunch of people. I felt good, and was release in a couple of days without any signs of gangrene except damaged muscle tissue. That led to a beer-belly without much beer.

It seemed like magic, but I knew that there is no magic, there is only technology. I also knew then, that the technology was not in that hospital.

I look forward to going back, where time has no meaning and beings can appear as they wish and do as they wish, forever.

* * *

So why did I write this book ? I wrote because it was always necessary. People need knowledge. People need truth. Good, bad, or ugly, we are entitled to know what is happening on our Earth. That statement especially applies to things that might affect us and our way of life. We need to know.

There are a number of reasons to get this information more visible:

A – You need to know it soon if you didn't already.

B – We Earthlings need to improve in many ways.

C – Hopefully, there are clues as to how we might improve.

D – And I apologize if the religious parts offend some, but until we know a bunch more about 'creation', these are the best guesses available now, IMHO.

E – Clearly, eventually, we all shall have no choice but to realize what you have just read is quite real. I do hope that you liked it.

Go research ! It's fun !

End of chapter 8

The Big Bang actually happened !   Page 63
Our Universe is ~13.7 billion years old !   Page 63
Creation as in Genesis !   Page 64
Evolution is a fact !   Page 64
Evil exists !   Page 65
God exists !   Page 65
Jesus is a real figure in history !   Page 65
Heaven exists !   Page 66
There is no Hell !   Page 67
Extraterrestrials are actually on Earth now !   Page 67
Extraterrestrials are peaceful !   Page 67
We will become interstellar space travelers !   Page 68
Our star, Sol, will explode and kill us all !   Page 69
The author's stories and writings are truthful !   Page 69

The Big Bang actually happened !

TRUE ! The majority of the astronomical community accepts the theory as fact. There are a couple of other theories, but they don't seem sensible. The following is from Space.com:

***

Simply put, it says the universe as we know it started with an infinitely hot and dense single point that inflated and stretched, first at unimaginable speeds, and then at a more measurable rate, over the next 13.7 billion years to the still-expanding cosmos that we know today.

Existing technology doesn't yet allow astronomers to literally peer back at the universe's birth, much of what we understand about the Big Bang comes from mathematical formulas and models. Astronomers can, however, see the "echo" of the expansion through a phenomenon known as the cosmic microwave background.

***

Our Universe is ~13.7 billion years old !

TRUE ! One method of measurement is stars in Globular Clusters. The method is a bit complicated. It involves the lifetimes of stars of different sizes and the fact that globular clusters of stars all have the (nearly) same birthday. For those astrophysicists, here is a good starting point:

Another method is using great space telescopes. The following is courtesy of NASA:

* * *

Update (July 21, 2023) ! Measurements made by <u>NASA's</u> WMAP spacecraft have shown that the universe is 13.77 billion years plus or minus 0.059. The age was further refined by ESA's Planck spacecraft to be 13.8 billion years old. They were able to do this by making detailed observations of the fluctuations in the cosmic microwave background and using that information in Einstein's Theory of General Relativity to 'run the clock backwards to time equal zero'.

* * *

Creation as in Genesis !

TRUE ! Given that physics allows a form of time travel due to warps in the space-time continuum. Also, the physics fact that time slows as speed increases, making time relative to one's location and movements. Either of these two facts can easily explain the six-day creation as in Genesis.

Even for those who insist upon 'literal' as printed, evolution is quite literal. The time as measured from our Earth is correctly 13.7 billion years. Our God's time is from a different perspective, and a different place, yet they are the same due to different clock speeds. This clock-speed difference due to different speeds is a proven fact, as discussed in chapter eight.

Evolution is a fact !

TRUE ! The evidence is overwhelming in quantity, quality, and branch of science. Radioactivity of elements, skeletons, bones, stone tools, age of rocks, age of our universe, plant remains-- the list is huge. Everything matches. Evolution happened over a very long period of **_our_** time.

Furthermore, it is logical, and still ongoing. Consider one simple example: Dogs-- man's best friend. In the recent past, there were AKC registered dog types that did not exist twenty years ago. Now there are types that nobody heard of just a few years ago. In the not-too-distant past, (say a few hundred years back) there were only a few types (breeds ?) of dogs. How did that happen ? Selective breeding and cross breeding is the how. Evolution is simply natural selection of types due to survival-ability and/or preference pressures, or offspring numbers/health.

Even today, there are many more people who are taller and broader than

80 years ago. Lots of them are fatter, too, but that might not be evolution. Two hundred years ago, people were even smaller on average.

Are we also getting more intelligent ? Hmmm... Maybe less ? Hmmm...

Evil exists !

TRUE ! None of us is perfect. From the very first time a child discovers that stepping on a cat's tail makes it 'yowl', and a tiny thrill goes through the child's body... Evil is present and responsible for that tiny thrill. A few people enjoy that thrill enough to continue and to escalate to much worse actions.

God knew all about Evil, which is why he gave us the original 'Ten Commandments', which seems to be a basic list of evil actions to avoid, such as murder. If evil did not exist, would God have a need to forgive our sins ? Would Jesus be needed ?

Life has always been a struggle of Good versus Evil. Evil has enough power to convince some very large religious groups to incorporate evil-doings into their religion and call it Good ! How such ideas took root is beyond comprehension, unless Evil power caused the implant.

True Good exists also, of course. It is stronger than Evil. Good is built-in and natural. Otherwise, we could not have evolved and multiplied and prospered. In the end, Good will prevail over Evil ! We should help in that epic, constant, global fight.

God exists !

TRUE ! Again, the evidence is overwhelming in quantity and quality. Also, many people have had experiences which seem to require an unusually Good power. Please re-read '**Why are we here ?'** in chapter eight. As for non-believers, they will eventually believe.

Jesus is a real figure in history !

TRUE ! Not only his factual existence is true as a person who lived in the years from near year zero through about the year 0035, he was very influential in the beginnings of all Christian religions and their offshoots. Jesus was probably somewhere between the ages of 33 and 39 when he died. It is difficult to determine the exact dates of birth and death for this famous person. An extremely interesting and thorough case is presented here:

https://www.bartehrman.com/how-long-did-jesus-live/

As for the truth of his Godly conception and power, there is a massive amount of writings, from 2000 years ago to the present. It is not difficult to believe. The difficult part is wondering which of the many denominations is correct or even most correct.

It seems that most of his teachings are not complicated to understand, to believe, or to follow. They certainly are not harmful for anyone. The various rituals and complications and rules that are required by the many various denominations actually weaken the case for Christianity in general. Even so, the basic belief that **he had God-like powers is probably true,** given the written evidence of those who knew him.

Heaven exists !

TRUE ! This author has 'stepped on the edge of Heaven', figuratively, and many others have as well. Its name sounds like 'haven'. Is that a coincidence ? In modern literature, there are plenty of stories similar to the one in this book. Just look up "NDEs". Here is one example (from NBC News) out of thousands around the world each year:

* * *

Dr. Melinda Greer, 65, was being evaluated for chest pain at a cardiac intensive care unit when her heart stopped. Dr. G., a retired pediatrician in Oklahoma, had asystole, a failure of the heart's electrical system which causes the heart to stop pumping, or flat-line.

That was 10 years ago. She is finally opening up about what she feels was a positive experience.

As the nurse was performing CPR on her, Dr. G. saw an "incredible white light" and felt "an incredible all-encompassing, all-surrounding sense of love."

She felt like she had returned to a "place that felt like home to me, and I was back among a group of what I can only call beings, because we weren't physical, that I considered my group."

It was "a wonderful experience," she said. "I really was angry when they brought me back."

After Dr. G. left the hospital, she decided to retire early, focusing on creative pursuits and new experiences, rather than acquiring things. She encourages people to get more involved in the "positive aspects of living in a beautiful world."

"Feel the wind, get out in nature, take off your shoes and socks and put your feet firmly on the ground and just listen to that inner voice, that's what I would recommend," she said. "I wish I'd done it long ago."

* * *

The experience described above is very typical of so-called NDEs. This author's experience was quite similar. Family and friends say that my personality changed greatly, and all in a very good way (It wasn't all that bad before, I claim).

That 'place' that she described, and my own description as well, must be Heaven ! There was no experience before or since that felt so wonderful and so full of love.

There is no Hell !

FALSE !! Researchers say that about 40% of NDEs are people who traveled to a demonic, Hellish place. Ancient writings say there is a Hell, and that Satan exists, and that other demons exist. This author does not wish to explore this subject further.

Extraterrestrials are actually on Earth now !

TRUE ! On November 13, 2024 the U. S. House of Representatives, during a hearing about UFOs and UAPs, etc., received a few answers that admitted knowledge of extraterrestrial presence and their vehicles, that go back many years. So, it is now official, in spite of the usual canned responses and lies from governments around the world.

The United States of America, as well as many other governments know these facts, but have kept them secret. They have lied, misled, produced fakes to ridicule the idea of extraterrestrials. They have ridiculed people who reported sightings, hid data, and generally made the whole idea seem silly. But it is not. It is real.

Meanwhile, the United States government has teams of experts working on crashed vehicles to reverse-engineer alien technology. Some of that technology is in various devices that are now being used (unknowingly) by we Earthlings !

One glaring example is computers. In the early 1960s, huge calculating machines (later to be called computers or laptops-- even phones) were only at a few universities or National Laboratories. They used thousands of radio tubes, required monster power supplies, and very large cooling structures. These calculating machines were not very reliable, but much faster than a person with a pencil. However, their computing and logic ability was micro-miniscule compared with a cheap, used laptop of today.

This remarkable change in compute-power came about from the invention of transistors (much smaller, cooler, faster, more reliable, and used less power), followed by silicon chip technology, which shortened the connection wires greatly. That shortened the time of electricity traveling to communicate between the transistors. Remember 'c', the speed of light ? It really matters !

Extraterrestrials are peaceful !

UNCERTAIN BUT LEANING TRUE ! There are known methods to completely wipe out the human population of Earth, even with our own older technology. Extraterrestrials, with much better technology and obviously more knowledge of physics than we possess, could easily employ a method that would empty our cool, green Earth. They could even make it painless, if they wished to kill us off. They are here. Why have they not commuted genocide yet ? They don't want to.

Or, if extraterrestrials wanted to simply take over control of Earth and use us as slave laborers, they could have done that long ago as well. They have not done so as yet. In other words, they won't.

It seems that they mostly hide (but not always, and not completely) and peacefully observe us. Most sightings have been above or near our military or scientific bases. Probably not for technology theft ! Probably not due to fear of our feeble weapon technology. Most likely to see how our technology is progressing, and most probably wishing that we would grow up enough to no longer need a military. Considering our history and current wars, they are probably trying to decide if we are (yet) a worthy addition to the galactic community. See chapter five.

We will become interstellar space travelers !

TRUE ! Of course, eventually we will have manned interstellar space vehicles if we can avoid destroying ourselves. The real question is when, as discussed in chapter five. It is difficult to imagine that we would forever forego the opportunity to explore other places, other worlds. Humans have always been explorers on the Earth. Vikings were famous for early human exploration. Columbus and others were driven to see what was beyond their sight.

Off this Earth, we have explored our moon somewhat, back in 1969. We have sent many rockets to many places, from asteroids to way out past Neptune and Pluto. We have sent robotic explorers to the surface of Mars, with a manned vehicle to follow soon. See NASA's plans for a 2030s voyage:

https://www.nasa.gov/humans-in-space/humans-to-mars/

Elon Musk has even earlier plans for his Spacex Company to set some human feet on Mars by about 2030 or sooner. Preliminary Spacex robotic spaceship voyages will begin very soon, with manned vehicles to follow:

https://www.space.com/spacex-starship-mars-launches-2026-elon-musk

Neither of these goals are starship interstellar trips; they are solar system trips. The difficulty of interstellar voyages is, as you now know, an order of magnitude more difficult with our current technology. If the extraterrestrials choose to include us as friends, we might become interstellar travelers much sooner. They may teach us their stardrive physics.

There is one more excellent reason to believe that we will eventually become interstellar travelers. Simply because we will have no other choice. See the following truth:

Our star, Sol, will explode and kill us all !

TRUE ! We don't have to worry about it yet, but at some point in our future, Sol will run out of hydrogen fuel and collapse. Then it will explode and instantly cook everything in the Solar System. The expected remaining lifetime of old Sol is around 5 billion years.

However, no one knows much about what happens during the millions of Sol's last billion or so years. Chaos might become rampant long before then. We will have to move as a massive, planetary societal group. Certainly, real climate change is in our future, not the natural fluctuations that Earth has always experienced since its slow cooling off period following its birth.

By the way, speaking of exploding, dying stars, see this website for a fascinating description of star death:

https://science.howstuffworks.com/star6.htm

When a star that is larger than our Sol, it dies in some very strange stages, which eventually results in lighter fuels (hydrogen, helium. Carbon, etc.) combining to become the heavier metals. Iron is one of them, but **iron can only be produced by a supernova** (a massive explosion of a type of star). The explosion then spreads the bits of the dead star far and wide around where the supernova was. Some of those bits and pieces eventually collect or are collected by far distant planets.

Earth has collected some of those chunks over its lifetime. Any iron of any sort that exists anywhere came from a dead supernova star ! If you have iron in your blood, it came from a very special huge star that died a horrible death long, long ago ! **We are all star-children !!**

The author's stories and writings are truthful !

MOSTLY TRUE ! Most statements in this book are true, including my own personal experiences (there were more than the three I wrote about). Many of the statements are almost certain, some are extremely probable. Some are simply probable. The few remaining statements outside these categories are quite possibly true. **None** of these statements is **false** except those which are stated to be false.

End of chapter nine

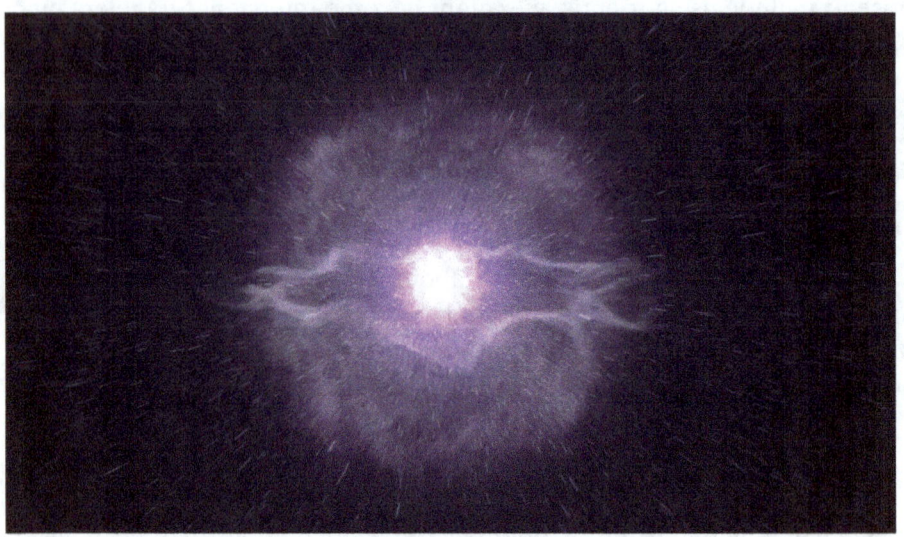

The Big Bang Begins !
Image courtesy NASA

**The author
and his
daughter !**

(Not really.)

**Laughter
is a good
tension
reliever !**

Other books by this author can be found at:

www.amazon.com/author/edmundjgoodwinsbooks

Make some notes about further questions here:

Make some notes about further questions here:

Make some notes about further questions here:

Make some notes about further questions here: